The Promise of Low Dose
Naltrexone Therapy

The Promise of Low Dose Naltrexone Therapy

Potential Benefits in Cancer, Autoimmune, Neurological and Infectious Disorders

ELAINE A. MOORE
and SAMANTHA WILKINSON

Foreword by Yash Pal Agrawal, M.D., Ph.D.

McFarland & Company, Inc., Publishers
Jefferson, North Carolina, and London

LIBRARY OF CONGRESS CATALOGUING-IN-PUBLICATION DATA

Moore, Elaine A., 1948–
    The promise of low dose naltrexone therapy : potential
benefits in cancer, autoimmune, neurological and infectious
disorders / Elaine A. Moore and Samantha Wilkinson; foreword
by Yash Pal Agrawal, M.D., Ph.D.
        p.   cm.
    Includes bibliographical references and index.

    ISBN 978-0-7864-3715-3
    softcover : 50# alkaline paper ∞

    1. Naltrexone.   I. Wilkinson, Samantha, 1965–   II. Title.
[DNLM: 1. Naltrexone—therapeutic use.   2. Dose-Response
Relationship, Drug. QV 89 M821   2009]
    RM328.M66   2009
    615'.19—dc22   2008048420

British Library cataloguing data are available

Cover art: Naltrexone chemical structure (wikimedia commons.com)

Manufactured in the United States of America

McFarland & Company, Inc., Publishers
    Box 611, Jefferson, North Carolina 28640
    www.mcfarlandpub.com

The authors wish to dedicate this book to their husbands, Rick Moore and Doug Flomer, and to the doctors and scientists, especially Dr. Ian Zagon, whose research efforts have revealed the promising potential of low dose naltrexone (LDN).

The information in this book is intended to educate patients and provide information about the use of low dose naltrexone. It is not intended as a substitute for medical advice. LDN should be used only under the direction and management of one's physician.

Acknowledgments

We have many people to thank for making this book possible. Never did we dream when we started the project how complex it would become and how many kind and generous people we would meet along the way. First and foremost, we'd like to thank our dear friend Dr. Judy Canfield for proofreading multiple drafts, offering invaluable suggestions for improvement, coordinating phone conferences with other physicians and scientific researchers, and being the first to introduce Elaine to low dose naltrexone and mention that a book on LDN was needed. Whenever we faltered, Judy steered us back on track.

Our warmest thanks also go to Dr. Ian Zagon, Ph.D., at Pennsylvania State University, for teaching us the ways in which LDN works and sharing so much of his valuable time speaking with us and replying to our questions. Thanks also to Dr. Jarred Younger, Ph.D., at Stanford University and Dr. Jau-Shyong Hong, Ph.D., at the National Institutes of Health, for answering our questions, sending us publications, reviewing early drafts, and teaching us about the intricacies of inflammation.

We'd also like to express our gratitude to our clinical-chemist friend Marvin G. Miller for creating the perfect illustrations to help us tell the story of LDN.

We are also indebted to Dr. Yash Pal Agrawal for meeting with us and sharing our enthusiasm while answering our many questions. We'd also like to take this opportunity to thank him for taking the time from his busy schedule to write the foreword for our book. We would also like to thank Dr. Bernard Bihari and his office staff for answering our questions and teaching us about the history of LDN; Dr. Ken Singleton for teaching us about the Lyme disease connection; Dr. Jill Smith for sharing pictures and information from her Crohn's disease trial; Dr. Maira Gironi for answering our questions about MS, and Dr. Keith Bodrero, DO, for taking the time to share his practical experiences with LDN. Our thanks also goes out to Dr. Skip Lenz, PharmD, for sharing his survey results and proofreading early drafts of our work and Bill Willis, FIACP, at Grandpa's Compounding Pharmacy in Placerville, California, for taking the time to talk with us and answer our many questions.

We'd also like to thank Dr. Bruce Cree, Dr. Burton Berkson, and Dr.

Jaquelyn McCandless for their help and support, and we'd like to thank Moshe Rogosnitzky from MedInsight for proofreading early drafts. Many patients were also helpful in sharing their LDN experiences, and we'd especially like to thank Tony White, Vicky Vinlayson, and Destiny Marquez. And thanks to Mary Bradley for all she has done for the LDN cause.

In addition, we'd like to thank our families, Rick, Brett, and Lisa Moore, and Doug Flomer, for supporting us every step of the way. And we can't forget to mention the World Wide Web. Not only has the Web brought LDN to the forefront of medicine, it's the common link that united everyone who worked with us on this project.

Table of Contents

Foreword
by Yash Pal Agrawal, M.D., Ph.D.

Snake oil or wonder drug? That is the question. When Elaine Moore asked me to write a foreword for her book, I jumped at the opportunity. The LDN (low dose naltrexone) story needs to be told, and I am happy to get the ball rolling. I first learned about LDN through an Internet chat board while investigating a temporary hearing loss that I had suffered.

At that time the dot.com bubble had just burst, and I was starting my first job as a young pathologist-investigator at the University of Iowa. Barry Marshall, the Australian pathologist (and future Nobel laureate), was still in the news regarding his discovery that a lowly bacterium (*H. pylori*) caused gastric ulcer and could be treated by antibiotics. Such a conclusion was of course anathema to Big Pharma, because Marshall's findings suggested that gastric ulcers could be treated with a couple of courses of antibiotics, as opposed to the lifelong anti-ulcer medication that Big Pharma preferred. This was heady stuff, since pathologists had been peering down microscopes for ages, many had seen those bacteria, but none had made the correlation that *H. pylori* caused gastric ulcers.

So with this background in mind, when I read on the internet that LDN prevented relapses in multiple sclerosis, I was intrigued, to say the least. Could it be that the inexpensive naltrexone, an FDA-approved drug that is used in much higher doses for other medical conditions, might be the silver bullet that everyone was seeking? From what I could gather from the Internet, the standard MS medications were not held in great regard by patients, mostly due to a perceived lack of efficacy, high cost and potential side-effects. Thus began my search for the truth about LDN and MS.

Very soon, it was obvious that LDN was reported to benefit almost every condition under the sun. As for MS, it was claimed to have a near perfect record in preventing relapses. Since I am a scientist by training, all of this smacked of a new snake oil remedy. However, the personal stories, which many patients cited, seemed genuine enough.

One such person I met on the Web and later in person was Samantha Wilkinson (www.ldners.org). She'd had advanced MS for many years, had undergone all kinds of standard therapy, including chemotherapy, and nothing seemed to work. She was going downhill and, in her own words, was at the end of her rope, until she found LDN and experienced a remarkable improvement. Hers and other reports from the United States and Great Britain (www.ldnresearchtrust.org) convinced me that all these people could not be exaggerating. After all, they had nothing to gain from it.

The problem was that no neurologist was willing to take these reports seriously and most just pooh-poohed them away. The problem also lay in the fact that there was nothing in the peer-reviewed literature to suggest that LDN worked. Any neurologist wanting to prescribe LDN would face condemnation from his or her peers and from Big Pharma. Attempts to get support from the National Multiple Sclerosis Society (NMSS) did not bear fruit either, or more precisely not the kind that patients were looking for. The NMSS issued a statement condemning LDN.

At this time I joined forces with Samantha and other Internet friends, Rob Lester and Art Mellor, to get at the truth regarding LDN and MS. Samantha was a computer professional, and one of the first things we did was a Web-based international survey to see if patients using LDN experienced benefits. The answer was a resounding yes. The next steps were harder, as I wrote to dozens of scientists in Europe and in the United States, urging them to do a clinical trial of LDN in MS. No one was interested, except for Dr. Maira Gironi in Italy, who at the time, did not have the funds to do the study.

We quickly realized that we would make no headway until there was something in the peer-reviewed scientific literature. The problem was that Dr. Bernard Bihari who had discovered the LDN effect, did not have patient records in a format that would bear scientific scrutiny and was otherwise unwilling to spare the time to document these cases. I myself lacked specialized training in neurology and the resources to conduct a clinical trial, so I did the next best thing.

Based on the available peer-reviewed literature, I constructed a medical hypothesis, indicating why I thought that LDN works in treating MS. Once this was published in the journal *Medical Hypotheses*, the LDN-in-MS story gained some credibility. Rob Lester helped further by posting my interview on Art Mellor's educational MS site (www.acceleratedcure.org). Mellor also guided our efforts to raise money for a clinical trial.

All of these efforts allowed us to partially fund a University of California, San Francisco investigator, Dr. Bruce Cree, to study LDN in MS. This was followed by other studies of LDN at Stanford and at other universities. The NMSS issued a more moderate statement regarding LDN. After many years, we had achieved the goal that we had set for ourselves, to bring LDN to the notice of mainstream neurologists and scientists, and to initiate clinical trials. While there is still little published proof that LDN works in MS, the genie is out of the bottle, and it is only a matter of time before the issue is resolved.

At this time it is appropriate to acknowledge and commend Elaine Moore for piecing together this remarkable story. While my comments pertain only to LDN and MS, the original claim that LDN is useful in a variety of conditions may not be so far-fetched after all. If LDN works by preventing oxidative damage—and oxidative damage is central to many disease conditions—then it is conceivable that LDN could be effective in a variety of disorders. This and other claims in the book have been and will likely continue to be verified by appropriate clinical studies and trials.

Dr. Yash Pal Agrawal is the director of the central laboratory at New York Presbyterian Hospital and an associate professor of clinical pathology and laboratory medicine at Cornell University.

Preface

Naltrexone is an opiate antagonist drug developed in the 1970s and approved by the United States Food and Drug Administration (FDA) in 1984 as a safe and effective treatment for opiate and alcohol abuse. Used at much lower doses in a protocol referred to as low dose naltrexone (LDN), naltrexone is reported to offer benefits in a wide range of diverse conditions including Parkinson's disease, autism, multiple sclerosis (MS), Alzheimer's disease, HIV infection and other viral illnesses, as well as several types of cancer and various autoimmune disorders.

In this book we describe the history of naltrexone, its pharmacological properties, its immune system effects, its potential therapeutic uses, research studies, clinical trials, and the reasons behind its large, worldwide patient following.

As a medical technologist primarily working in toxicology, and an autoimmune disease patient, I've followed the story of LDN and the results of its few clinical trials with great interest. As a patient with multiple sclerosis (MS) who has personally benefited from LDN for the past several years, my coauthor Samantha Wilkinson is a passionate and realistic advocate of its use. As a computer analyst, Samantha has developed an ongoing survey that captures the patient experience. She presented the initial data from her survey at the first annual LDN conference in New York City in April 2005.

Many anecdotal reports related to the use of LDN can be found on the Internet. In her book *Up the Creek with a Paddle,* Mary Bradley describes her quest to obtain LDN for her husband who has used it successfully for his progressive form of MS. The results of several clinical trials using low doses of opiate antagonists have been published, and some remarkable research in this area continues at Pennsylvania State University's College of Medicine by Doctors Ian Zagon and Jill Smith, at Stanford University by Dr. Jarred Younger, and at the National Institutes of Health by Dr. Jau Shyong Hong. However, little information is available that describes exactly what LDN is or that explains in lay terms how it works in the body. And unfortunately various misconceptions about LDN and its use have been perpetuated over the years.

Our primary goal in writing this book was to acquaint readers with the potential therapeutic benefits of LDN in disorders for which it has been used

5

or studied and to explain what can be expected from its use. Along with chapters on the history of low dose naltrexone and its applications in specific disorders, we've included a practical chapter for those interested in its use, explaining how LDN is administered, including information on fillers and compounding pharmacies along with a list of doctors who prescribe LDN, a chapter listing available resources, an appendix listing clinical trials, and a glossary of terms.

Our secondary goal was to generate interest in the need for more extensive clinical trials and to encourage funding in this area. The results of several past trials have been hindered by poor study design and a lack of adequate funding. And while several clinical trials have demonstrated the potential of LDN, more trials are needed to confirm and expand upon these results. Without FDA approval acknowledging the safety and efficacy of LDN in specific disorders, its true potential remains unexplored. Consequently, patients who might be able to benefit from LDN, especially patients with conditions lacking effective treatments or who have difficulty obtaining prescriptions for LDN, are deprived of any potential benefits.

To achieve our goals, we've grounded our work in available clinical and scientific research, with an emphasis on the effects of LDN in restoring homeostasis. We've also interviewed the country's top researchers and learned the reason why they persist in studying LDN despite limited funding opportunities and a lack of support from pharmaceutical companies. Lastly, our work would not be complete without input from patients who describe the importance of LDN in terms of its ability to provide relief from pain, halt disease progression, and facilitate the body's ability to heal.

Our story would also not be complete without an underlying thread describing the influences of the Internet. Without the power of the Internet, the story of LDN would be confined to a handful of university research laboratories. Consequently, very few people would have heard of Dr. Ian Zagon or Dr. Bernard Bihari, and the thousand of patients who are using LDN today would have missed out on the opportunity to experience its benefits firsthand. For this reason we've included information on the unprecedented influences of the Internet and the ways it can both empower and mislead patients.

—Elaine A. Moore

Introduction

The drug naltrexone was developed in the 1970s in response to the need for an effective treatment for heroin addiction. Shortly after its development, doctors Ian Zagon and Patricia McLaughlin at Hershey Medical Center, Penn State University College of Medicine, began studying naltrexone to see if it might offer benefits in other diseases.

In 1982, Zagon and McLaughlin discovered that naltrexone affects cell growth in different ways depending on whether high or low doses of the compound are used. This prompted Zagon and his team to pursue studies of naltrexone in treating various medical conditions. Coincidentally, these researchers also discovered that naltrexone triggers increased production of naturally occurring opioid compounds known as endorphins when used at a fraction of the usual dose.

In the mid–1980s the Harvard-educated, New York City neurologist Bernard Bihari, director of the Division of Alcoholism and Drug Dependency at the State University of New York (SUNY) Health Service Center at Brooklyn, began using naltrexone for drug addicts under his care. Because many of these patients were also infected with the human immunodeficiency virus (HIV), which causes acquired immune deficiency syndrome (AIDS), Bihari was able to evaluate the effects of naltrexone in this disorder. Bihari noted a reduction in the incidence of Kaposi's sarcoma in these patients that he attributed to naltrexone, which was approved by the United States Food and Drug Administration (FDA) in 1984 as a treatment for opiate addiction.

Encouraged by personal observations of naltrexone as well as the findings of Zagon and McLaughlin, in 1985 through 1986 at the Downstate Medical Center in New York, Bihari conducted the first clinical trial for low dose naltrexone (LDN) in patients with HIV infection. The trial was funded by the Foundation for Integrative Research. During the trial, Bihari demonstrated with laboratory tests that LDN appeared to improve immune function in HIV infected patients. He presented a report describing these findings at the Second International Conference on AIDS in Paris, France, in June 1986.

Upon completion of the trial, Bihari returned to private practice and began prescribing LDN for patients with HIV infection and various other immune-mediated diseases, including multiple sclerosis and lymphoma. Since then, Bihari has treated hundreds of patients suffering from a wide range of unrelated illnesses. Bihari's childhood friend, the physician David Gluck, has brought attention to Bihari's findings on an informational Web site.

As a result, thousands of patients worldwide, particularly HIV, cancer, and MS patients, have requested prescriptions for LDN, and hundreds of physicians have incorporated the use of LDN into their treatment protocols. According to informal patient surveys, pharmacy surveys, and anecdotal accounts, the majority of these patients have benefited from its use. However, informal surveys and anecdotal accounts are incongruous with solid scientific research. Fortunately for patients worldwide, Ian Zagon and his team at the University of Pennsylvania's College of Medicine have pursued their study of naltrexone and other opiate antagonists for the past twenty-five years.

While researching LDN, Zagon discovered that its primary effect is increased production of the endogenous opiate met-5-enkephalin, which he named "opioid growth factor" (OGF) in reference to its functional properties. Zagon's studies of the opioid growth factor and opioid growth factor receptor led to the discovery of OGF as a therapeutic agent.

Along with Dr. Jill Smith, Zagon conducted a pilot clinical trial of LDN in Crohn's disease that resulted in benefits for 89 percent of the subjects who received the drug. In clinical trials of naltrexone and OGF, Zagon has discovered improvement in wound healing in patients with diabetes, and he has demonstrated improvement in patients with pancreatic cancer as well as in patients with tumors of the head and neck. Other researchers have also spent years studying the potential therapeutic benefits of opiate antagonists. At the University of Iowa and later at Cornell University, Dr. Yash Agrawal researched and wrote about the role of LDN in reducing oxidative stress and inhibiting the production of excitotoxins that contribute to neurodegenerative diseases (*Medical Hypotheses* 64, no. 4 [2005] 721–724).

However, because formal clinical trials and publications are few, and FDA approval for using naltrexone in these other medical conditions lacking, LDN has not been embraced by mainstream medicine. Physicians aren't encouraged by pharmaceutical companies to prescribe LDN. Medical journals don't run full-page ads advocating its use. Furthermore, pharmaceutical companies have little incentive for funding clinical trials on a generic drug that has already been FDA approved and widely manufactured as a 50-mg tablet. Consequently, LDN, which is used in doses ranging from 1.5 to 10 mg, must be compounded by pharmacists into smaller doses or diluted before it can be used.

Despite the lack of advertising, the scarcity of funding for clinical trials, and the reluctance of many conventional doctors to prescribe it, LDN continues to grow in popularity. The first annual conference for LDN in 2005 attracted physicians and patients from across the globe and each successive conference has attracted a wider following.

In particular, the success of LDN in MS has garnered tremendous attention. Across the globe, a legion of Web sites created by both physicians and patients describe protocols for the use of LDN in MS and detail successful accounts of its use. Since 2005, publications as diverse as the *Townsend Newsletter for Doctors*, the *Boston Cure Project Newsletter*, the *Fort Meyers Florida Island Newsletter*, *Medical Hypotheses*, and the *London Herald* have published articles on LDN stating the urgent need for clinical trials.

On April 20, 2007, the National Cancer Institute jumped on the bandwagon and hosted a conference in Bethesda, Maryland, "Low Dose Opioid Blockers, Endorphins and Metenkephalins: Promising Compounds for Unmet Medical Needs," that focused on the potential therapeutic benefits of naltrexone and related opiate antagonists. Here, researchers described several ongoing clinical trials of LDN in multiple sclerosis, pancreatic cancer, and Crohn's disease.

Dr. Nicholas Plotnikoff from the College of Pharmacy at the University of Illinois at Chicago presented a paper he wrote with Dr. Bernard Bihari describing a clinical trial of metenkephalin in patients with AIDS-related complex (ARC). At the conference, several doctors, including Burton Berkson, also described their successful use of LDN in cancer and Parkinson's disease.

The year 2007 was a banner one for LDN. The first medical journal article published in the United States describing LDN's efficacy in treating Crohn's disease appeared in the January 2007 online edition of the *American Journal of Gastroenterology*. In October, researchers and patients gathered at the third annual LDN Conference in Nashville to present LDN progress reports and survey data. Also in late 2006 and 2007, clinical trials of LDN in MS were held in Germany, San Francisco and Italy; a phase 2 trial of Opioid Growth Factor in pancreatic cancer was in progress at Pennsylvania State University; and a pilot clinical trial of LDN in fibromyalgia began at Stanford University. In addition, the Mali clinical trial of LDN in HIV/AIDS commenced in December 2007.

Hardly a magic bullet, LDN is reported to halt disease progression in a number of different conditions, and in some cases it has improved symptoms remarkably. However, LDN doesn't offer benefits in all patients who try it. And while side effects appear to be rare and mild when used appropriately, clinical trials are necessary to evaluate potential adverse effects and determine the true efficacy and safety of LDN in specific conditions.

Why some patients experience benefits and others don't is a topic worth exploring and something that clinical trials and further studies may help determine. In this book our goal is to acquaint readers with LDN's potential to restore homeostasis and improve health while emphasizing the genuine need for more extensive clinical trials.

1

The War on Drugs—
A History of Naltrexone

Naltrexone was initially developed as a strategic treatment intended to combat heroin addiction. Coincidentally, researchers at Pennsylvania State University studying the smaller birth weight of infants born to heroin addicts serendipitously discovered that, used in low doses, naltrexone and related compounds affect cell growth, healing, and immune function. These observations were the first hints of naltrexone's potential benefits in medical conditions having no relationship whatsoever to drug addiction. This chapter describes the development of naltrexone and explains how the uses of low dose naltrexone (LDN) came to light.

1960s America

The 1960s were a paradox, with the best of times and the worst of times rapidly shifting gears. The assassination of President John Kennedy, the introduction of civil rights laws, the student massacre at Kent State, the military draft, and our country's involvement in the Vietnam War pulled its systems into social turmoil at home and military turmoil abroad.

In addition, the post–World War II chemical renaissance had blossomed into full swing. The pharmaceutical industry had escalated its efforts to calm, stimulate, and otherwise medicate the public while illegal drugs had a distinct appeal for people in all walks of life. For many individuals, a cycle of drug addiction escalated to the dead end of heroin addiction. Consequently, by 1970 the use of heroin both in the United States and among U.S. military troops serving abroad had reached epidemic proportions.[1]

Heroin, Opium and the Poppy Plant

Drugs such as heroin and morphine exert their effects by mimicking the body's naturally occurring pain-reducing substances, which are known as endogenous opioid peptides. Endogenous is a term meaning *from within* that refers to substances the body produces naturally. Peptides are chains of alpha amino acids linked by amide or peptide bonds. When five amino acid molecules are linked together, peptides are referred to as pentapeptides.

Endogenous opioid peptides, which function as neurotransmitters, neuromodulators and hormones, include the endorphins, enkephalins, and dynorphins. The endogenous opioid peptides are involved in the perception of pain, the modulation of behavior, and the regulation of autonomic and neuroendocrine function. Drugs such as morphine and heroin, with properties similar to those of the endogenous opioids, are derived from the poppy plant *Papaver somniferum*. The poppy plant is the main source of non-synthetic narcotics worldwide.

Opiate Terminology

The word *opium* is derived from *opos*, the Greek word for juice, and refers to the latex juice produced within the unripe seed pods of the poppy plant. Opium products and related plant alkaloids are referred to as opiates. The term *opioid* encompasses all compounds related to opium, including endogenous opioid peptides. Drugs isolated from the opium poppy or synthesized to resemble the exogenous (produced outside of the body) opioids include morphine, codeine, heroin, methadone, hydrocodone, and oxycodone.

Opiates depress nerve transmission in sensory pathways of the spinal cord and brain that signal pain. This makes opiates particularly effective painkillers. Opiates also inhibit brain centers that control coughing, breathing, and intestinal motility. For this reason, codeine is one of the most effective cough suppressants and opiates in general are considered respiratory system depressants. Opiates are also exceedingly addictive, quickly producing tolerance and dependence.

Opiate Use

Originating in Asia Minor as early as 300 BC, the poppy plant has long provided mankind with opiates used as both medicines and intoxicants. Records from the sixteenth century indicate that physicians in Arabia and Europe used opiate compounds for pain relief and anesthesia. In the early eighteenth century, when opiates were first made into a tincture called *paregoric*, Arabian traders supplied opium to the Chinese to treat dysentery.

In the nineteenth century, opiates were widely sold legally and inexpensively in the United States. Physicians prescribed opiates to their patients, while drugstores, grocery stores, general stores, and mail-order catalogs stocked and sold opiates to people without prescriptions. Patent medicines marketed for teething, diarrhea, women's troubles, and cough all contained opiates. Although poppies were grown in the United States and provided some opiates locally, most of the morphine available in the United States in the nineteenth century was imported legally.

Morphine and Heroin

In 1805, the German pharmacist Friedrich W. Serturner isolated the active ingredient in opium and named it *morphium* after Morpheus, the Greek god of dreams. This compound later became known as morphine. Morphine acts within the central nervous system, particularly at synapses of the nucleus accumbens within the forebrain, to relieve pain. In the management of pain, morphine is considered superior to other narcotic analgesics. Unfortunately, morphine is highly addictive compared to other analgesics. Physical and psychological tolerance to—and dependence on—morphine develop rapidly.

In the 1870s, chemists synthesized a supposedly nonaddictive substitute for morphine known as 3, 6-diacetylmorphine by acetylating (adding an acetyl group to) pure morphine. In 1898, the Bayer pharmaceutical company of Germany became the first manufacturer of this compound, which it named *heroin*. Heroin is two to three times more potent than morphine, primarily because of its increased lipid solubility, which provides accelerated central nervous system penetration. Early on, heroin was sold as a superior remedy for cough and marketed as a treatment for morphine addiction. It didn't take long for addicts to realize that heroin was much more potent when injected.

The Opiate Problem

By 1830, government officials in China recognized an association between illegal activities and opium addiction. They tried banning the importation of opium into China and a series of unsuccessful opium wars took place. Unfortunately, the defeat of China in the opium wars of 1842 and 1856 allowed for the free trade of opium into China as well as other countries. By 1910, the effects of the free opium trade were taking their toll on the U.S. population.

The Harrison Narcotics Act of 1914 was the first U.S. law that banned the domestic distribution of drugs. This act was passed to regulate the production and distribution of opiate-containing substances under the commerce clause of the U.S. Constitution.

Heroin in the 1940s and 1950s

After World War II, heroin and related opiates were widely embraced by artists and musicians as well as a more discreet cluster of physicians, pharmacists, and other medical professionals in the United States. Also enchanted were the poor people in America's slum areas including Spanish Harlem in New York City, where heroin use was reported to be epidemic in the mid–1950s.[2]

Although the possession and distribution of heroin were criminal offenses, the heroin trade was a lucrative business in the 1950s, and the Mafia was heavily involved in heroin trafficking.

The first period of large-scale heroin smuggling into the United States since its prohibition occurred from 1967 through 1971. Turkish opium was processed into heroin in France and then smuggled into New York in a clandestine operation known as the *French Connection.* In the mid–1970s Mexican brown heroin emerged. Sold at a lower price than European heroin, it became readily available in the West and Midwest. Suddenly, heroin was easy to obtain and affordable.

The Federal Role in the 1960s and 1970s

Although heroin addiction was recognized as a serious medical and social threat, treatment for heroin addiction in the 1960s was scarce. In New York City, addicts aged twenty-one or older were given the choice of prison or a stay at the Addiction Research Center in Lexington, Kentucky, which had a long waiting list. By 1971, both the executive and legislative branches of the government authorized funding for expanded research and treatment of opiate dependence.

THE WAR ON DRUGS. In 1971, President Richard Nixon announced the "War on Drugs," a prohibition campaign intended to reduce the supply and diminish the demand of illegal drugs. The campaign included laws and policies to discourage drug trafficking and consumption, as well as programs designed to treat drug abuse.

In June that year, Nixon created the Special Action Office for Drug Abuse Prevention (SAODP) and appointed Dr. Jerome Jaffe as its director. Jaffe's goal was to move drug-abuse treatment out of the prison and hospital settings and into community-based drug-rehabilitation services. Consequently, for Jaffe the development of a successful treatment for narcotics addiction became high priority, and he had the government resources to speed its clinical development.

In March 1972, Congress passed the Drug Abuse Office and Treatment Act, calling for the development of long-lasting, nonaddictive, blocking and antagonist drugs or other pharmacological substances for the treatment of heroin addiction. This act provided substantial financial support for research.

But Jaffe cautioned in the foreword he wrote for the National Institute of Drug Abuse Monograph 9, *Narcotic Antagonists: Naltrexone Progress Report,* published in September 1976, "Antagonists might ultimately prove to be only of value to a limited subgroup of the opiate population."[3] And this has turned out to be the case.

Opiate Antagonists and Naltrexone

Although the exact mechanisms regarding how antagonists worked hadn't yet been discovered, as early as the 1940s scientists knew that there were drugs, which they called antagonists, which blocked the physiological effect of other drugs as well as the body's natural or endogenous chemicals. For instance, beta adrenergic blockers or beta blockers were developed to reduce the adrenergic stimulating effects that led to hypertension.

In the foreword to NIDA's Monograph 9 mentioned above, Jaffe explains that the use of opiate antagonists to treat opiate addiction was first proposed by Abraham Wikler in 1955, when he postulated that opiate dependence caused a "synthetic" need that was readily satisfied by opioids. The withdrawal phenomena, Wikler proposed, could be conditioned to environmental stimuli, such as opiate antagonists that blocked the positive reinforcement of drug-seeking behavior.

The Search for a Narcotic Antagonist

With that aim, for the next decade researchers set out to develop a narcotic antagonist that might prevent death from narcotic overdoses and effectively treat narcotic addiction. At the Addiction Research Center in Kentucky, Dr. William Martin and his coworkers initiated a series of studies that showed that a narcotic antagonist could be effectively used to block the euphorigenic (euphoria-inducing) and dependence-producing properties of opioids in humans.

The relatively pure opiate antagonist naloxone was considered (well known in emergency rooms as Narcan, it was first synthesized in 1960), but its short duration of action and ineffectiveness when taken orally precluded its use in drug addiction. Naloxone is usually administered intravenously.

Choosing Naltrexone

A similar compound, naltrexone, which was first synthesized in 1963 and at the time still under development, seemed a better choice. An inexpensive compound, significantly more potent than naloxone, naltrexone can be administered orally, and it has a fairly long mode of action (twenty-four hours), which made it a potential candidate as a treatment for heroin addiction. Federal officials were very interested in expediting naltrexone's development.

Deciding on a Dose

Early trials of naltrexone, which are described later in this chapter, focused on safety first. Once the drug was found to be safe at a certain dose, this dose was used in clinical trials designed to study the drug's efficacy. Perhaps trials employing different doses would have provided some insight into the benefits naltrexone provides at low doses in other conditions.

However, toying with different naltrexone doses wasn't considered until the 1980s when the Pennsylvania State University researcher Ian Zagon found that at low doses naltrexone decreased cell proliferation and increased production of endogenous opioid peptides. Encouraged by these findings, the New York City neurologist Bernard Bihari began studying the effects of low dose naltrexone (LDN) on heroin addicts afflicted with symptoms of acquired immune deficiency syndrome (AIDS) or AIDS-related complex (ARC). The studies of doctors Zagon and Bihari are briefly described later in this chapter and more comprehensively in the chapters on specific diseases·

Toxicology and Drug Doses

Nearly 500 years ago, the German physician Paracelsus (1493–1541), who is often regarded as the father of toxicology, wrote, "All substances are poisons; there is none that is not a poison. The right dose differentiates a poison and a remedy." This is often condensed to "the dose makes the poison."[4] Paracelsus, who used opiates extensively in his practice, is also credited with compounding a tincture of opium, which he named laudanum. Today, laudanum remains an official preparation.

The relationship between a specific dose and its response or effect remains one of the most fundamental concepts of toxicology, which is the study of the adverse effects of chemicals on living organisms. A substance can produce harmful effects associated with its toxic properties only if it enters the body in a sufficiently high concentration. Even a substance as seemingly benign as water can be fatal if excessive amounts are ingested.

Therapeutic Drug Doses

Another key concept in toxicology is that the optimal therapeutic dose is the lowest dose of a drug needed to produce its desired effects (its efficacy). That is, for all chemicals there is a dose response curve, or a range of doses that results in a graded effect between the extreme of no effect and a complete response or toxic effect. The desired effect of the drug, in most cases, can be related to the blood level of the drug. For instance, a 50-mg daily dose of naltrexone is reported to show efficacy in preventing relapse in heroin addiction, and this dose usually results in a blood naltrexone level of approximately 2 ng/ml.

The therapeutic dose of a drug describes the dose in which the intended effects usually occur. Blood drug levels that fall below the therapeutic range are generally considered ineffective while levels above the range are associated with toxicity or lethality.

Pharmacology

Drugs have chemical properties that allow them to react with the cells of living organisms, causing specific effects. Drugs accomplish this by binding to specific receptors found on cell membranes. To achieve their intended effects, drugs must be absorbed by the body and then transported and distributed to certain sites in the body. Drugs also have a property known as fate. Fate determines how long drugs linger in the body before being transported to the liver, metabolized and excreted. Drugs also have a half-life, which is the time it takes for blood levels of the drug to be reduced by 50 percent.

Many factors, including body weight, other medications, diet, general health, liver function, and metabolism affect a specific drug's rate of absorption, as well as its fate and metabolism. Although a certain drug dose usually causes the intended actions and results in therapeutic blood levels, this isn't always the case. In practice, not everyone will respond to medications in the same way. A dose that works for most people may cause toxicity in others or it may have a negligible effect. This is important to keep in mind when assessing LDN for its ability to elicit improvement in so many different conditions, including AIDS, cancer, and neurodegenerative and autoimmune disorders.

Development of Naltrexone

While it was under development, naltrexone was known as Endo 1639A or EN-1639A. Naltrexone is a pure opiate antagonist with the chemical structure $C_{20}H_{23}NO_4$-HCL. This compound was first synthesized as a white powder in 1963 by Endo Laboratories, a small pharmaceutical company with extensive experience in narcotics, located in Long Island, New York. In 1969, the DuPont Company purchased Endo Laboratories and acquired the rights to naloxone and also Endo 1639A, which they later named naltrexone.

The Special Action Office for Drug Abuse Prevention (SAODP) created by President Nixon began to phase out of existence in mid–1974. Consequently, the task of developing narcotic antagonists fell to the newly formed National Institute on Drug Abuse (NIDA). NIDA officials, eager to find a narcotic antagonist and win the war on drugs, approached DuPont and offered their assistance in moving naltrexone swiftly through the FDA approval process. DuPont agreed to pursue the development of naltrexone, and NIDA agreed to pay the bulk of the development costs.

The Cost of Development

From 1973 to 1974, NIDA supported twenty-six grants and contracts in pre-clinical and clinical studies related to narcotic antagonists at a cost of more than five million dollars. Approximately seventeen of these grants dealt with the use of naltrexone in clinical situations for the treatment of opiate abuse. The remainder of the grants and contracts consisted of controlled and uncontrolled trials with naltrexone, with the 1974 studies representing phase II testing.[5]

Naltrexone's Clinical Trials

Preclinical trials established naltrexone's safety in rats, rabbits, dogs and monkeys at moderate doses (up to fifty times the dose used in humans, which is 1 mg/kg). When much higher doses were administered subcutaneously (300 mg/kg) in monkeys, the drug caused prostration and convulsions and proved to be lethal in all four animals tested.[6]

Human trials concurred that the drug was safe used orally in the recommended doses of 50–300 mg daily. Trials for determining the drug's efficacy were another matter. Patients had to remain drug free for five to ten days before participating in trials. Most heroin addicts had no intention of complying and no desire to be treated.

Naltrexone blocks the "high" effect associated with opiates although it doesn't block cravings, something another recently introduced drug for opiate addiction known as methadone, although itself addictive, was able to do. Thus, recruiting addicts for trials of naltrexone was difficult and, because naltrexone lacks the reinforcing effects of methadone, social support services and expensive counseling were required.

Naltrexone's FDA Approval

The results showed that naltrexone was moderately successful in reducing heroin use. In 1984, the FDA approved naltrexone as an orphan drug, which provided seven additional years of exclusive marketing for DuPont.[7]

Marketing naltrexone was another matter. The product label stated that naltrexone, initially sold as the patent drug Trexan, didn't reinforce medication compliance. In addition, desired effects (thwarting the desire for heroin) could only be expected when external circumstances supported the use of the medication. In other words, if the addict had no intention of quitting, naltrexone wouldn't help. The recent introduction of methadone, another drug substitute for heroin, also worked against naltrexone's success. Methadone clinics were reluctant to recommend naltrexone (and lose business), and most drug rehabilitation facilities couldn't afford to implement it into their programs. Sales for 1995 indicated that less than 5 percent of the country's heroin addicts were using Trexan.[8]

Naltrexone 1995–2008

With assistance from the National Institute on Alcohol Abuse and Alcoholism, the U.S. government funded additional clinical trials for naltrexone, now marketed as ReVia, for use as a treatment in alcoholism. To expedite the process, the FDA also linked phase IV clinical trial requirements to annual sales and allowed for flexible phase IV trials. As with the naltrexone trials for opiate abuse, patient compliance was poor and recruiting patients was difficult.

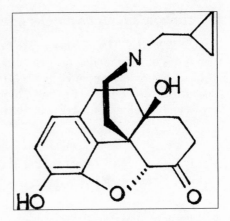

Naltrexone chemical structure (Marvin G. Miller).

Results showed naltrexone to be little better than a placebo for alcohol addiction unless it was administered as part of a multidisciplinary treatment program. With required labeling stating that ReVia should be considered only one factor in a comprehensive alcohol treatment package, naltrexone was approved in 1995 as a 50-mg daily dose for the treatment of alcoholism. As an approved alcohol therapy, the FDA granted DuPont three additional years of post-approval market exclusivity. In 1997, the patents expired and naltrexone became a generic drug.

Recent Naltrexone Initiatives

Since its approval for opiate and later alcohol addiction, Naltrexone has been studied extensively, primarily in NIDA-funded trials for everything from weight loss in smoking cessation to cocaine addiction· A search on the NIDA Web site alone brings up over 550 pages links mentioning naltrexone.[9] The latest studies have involved sustained-release depot doses used in an effort to increase compliance in narcotics abusers. Several studies tested low doses of naltrexone to see if LDN might work for smoking cessation and as an adjunct therapy for pain management used in conjunction with opiates to prevent opiate dependency.

Naltrexone hydrochloride is available in a number of different preparations, including Antaxone, Bristol-Myers Squibb brand of naltrexone hydrochloride, Celupan, DuPont brand of naltrexone hydrochloride, Lacer brand of naltrexone hydrochloride, Lamepro brand of naltrexone hydrochloride, Nalorex, Nemexin, Orphan brand of naltrexone hydrochloride, Pharmazam brand of naltrexone hydrochloride, ReVia, Schering-Plough brand of naltrexone hydrochloride, Trexan, United Drug brand of naltrexone hydrochloride and naltrexone hydrochloride.

Naltrexone's Uses as an Orphan Drug

In 1998, the FDA awarded orphan drug status to naltrexone for treating symptoms of childhood autism and as a therapy for self-injurious behaviors. Orphan drug status is granted by the FDA to qualifying products intended for diagnosis, prevention and treatment of rare diseases where no current therapy exists, and which affect fewer than 200,000 patients in the United States.[10]

Bernard Bihari and Low Dose Naltrexone

Bernard Bihari is a Harvard-educated neurologist. In the early 1980s, Bihari began using naltrexone as a treatment for drug addicts in New York City. He soon realized that addicts using the recommended 50–300-mg daily doses of naltrexone felt terrible since naltrexone blocked not only the opiate alkaloids in heroin but the patients' own endogenous endorphins, which contribute to well-being. The addicts complained of feeling miserable and not being able to sleep.

Around this time, a significant number of heroin addicts in the New York City area began to show signs of acquired immune deficiency syndrome (AIDS). In his drug-addiction clinic, Bihari noticed that drug addicts with AIDS using naltrexone for drug addiction were not developing signs of lymphoma and other symptoms of immune dysfunction. After reading the reports published by Ian Zagon and his team regarding low dose naltrexone's effects on cell proliferation and healing, Bihari began studying LDN in patients with conditions not responding to other therapies. After conducting the first clinical trial of LDN in AIDS, which is described in chapter 8, Bihari returned to private practice and began using LDN for patients infected with HIV and later in patients with multiple sclerosis, lymphoma, and other immune-mediated conditions.

How LDN Works

As an opiate antagonist, naltrexone blocks the opiate receptor, a protein molecule found on various cells, including immune-system cells. The opiate receptor can be compared to a lock that is opened by a key. As the only substances that can activate the opiate receptor, opiates and opiate antagonists are the keys. Naltrexone closes the opiate receptor lock and blocks other chemicals, including endogenous opioid peptides, from opening the lock. In low doses, naltrexone blocks the receptor for four to six hours.

Once the opiate receptor blockade wears off, there is a rebound effect or response in which the body increases its production of endogenous opioid peptides. The body's cells respond to these peptide neurotransmitters by initiating various biochemical and cellular changes that inhibit cell growth, promote healing, and reduce inflammation, thereby helping restore homeostasis. The effects of low dose naltrexone are directly related to its ability to increase levels of endogenous opioid compounds.

Homeostasis

Homeostasis in living organisms refers to a negative feedback system in which the organism regulates its internal environment so as to maintain a stable, constant condition of good health. Simply put, all of the body's cells and systems work together to maintain health. For instance, when the kidneys are damaged or injured and can no longer filter waste efficiently, blood nitrogen levels rise. In response, muscle tissue cells release a chemical known as lactic acid in their attempt to maintain a normal acid-base balance or pH. In addi-

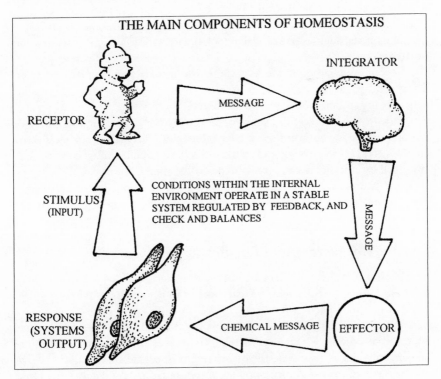

Homeostasis (Marvin G. Miller)

tion, white blood cells travel from the blood and lymph nodes to the affected area in an effort to prevent infection.

The concept of homeostasis originated in 1865 with Claude Bernard who is often considered the father of physiology. The term homeostasis was first used to describe this concept in 1932 by Walter Cannon and is derived from the Greek words *homos* for same and *stasis*, which refers to standing.

The Opiate Receptor

In 1972, Johns Hopkins graduate student Candace Pert and her colleagues discovered the elusive opiate receptor.[11] The opiate receptor, which had long been suspected of existing but had never been found before, is a small protein molecule that sits on the cell surface much like a keyhole. Although the opiate receptor was initially found on neurons or brain cells, it's now known that opiate receptors exist on immune system cells as well as on most tumor cells.

Substances known as opiate agonists are able to attach to the opiate receptor, acting as ligands and thereby activating the receptor. When the opiate receptor is activated, neurotransmitters or chemical messengers are released. These chemical peptides evoke biochemical changes that can relieve pain. Opiate antagonists, such as naltrexone, block opiate agonists from activating the opiate receptor.

Several subtypes of opiate receptors exist, primarily mu, kappa, and delta. These subtypes are described later in this chapter. Specific opiate agonists can activate only specific opiate receptor subtypes. For instance, morphine primarily activates the mu receptor, slightly reacts with the kappa receptor, and does not activate the delta receptor.[12] Each of the opiate receptor subtypes is associated with specific physiological effects. Similarly, opiate antagonists primarily block specific opiate receptor subtypes rather than all subtypes of the opiate receptor. Naltrexone preferentially blocks the mu opiate receptor. Naltrexone is also considered a pure opiate antagonist because it has no agonist properties.

Discovery of Endogenous Opiates

Within three years of Candace Pert's discovery, the Scottish researchers Dr. John Hughes and Dr. Hans Kosterlitz at the University of Aberdeen discovered naturally occurring hormone-like proteins with opiate-like properties, which they called endogenous opiates. In addition, they found that endogenous opiates, which were further classified as endorphins, dynorphins, and enkephalins, also activate the opiate receptor. All substances that activate the opiate receptor are known as opiate receptor agonists and substances that block the opiate receptor are known as opiate receptor antagonists.

Opiate Receptor Subtypes

The primary opiate receptor subtypes include mu (u1), mu (u2), delta, and kappa. Specific compounds activate each subtype and each receptor subtype is associated with its own biochemical effects. However, changes in the dose of opioid compounds and the duration of their action can change which receptor an agonist compound activates or an antagonist compound blocks. These changes are important in understanding how LDN exerts its effects, a topic described in chapter 2.

TABLE 1.1 PRIMARY OPIATE RECEPTOR SUBTYPES

Opiate Receptor Subtype	Primary Physiological Effects
Mu (u1)	Spinal and supraspinal analgesia
Mu (u2)	Respiratory depression, euphoria; vomiting; inhibition of gut motility; propensity for physical dependence
Delta	Spinal analgesia
Kappa	Spinal and supraspinal analgesia

Neurotransmitters Modulated by Opiates

Opiates have potent analgesic effects. They exert these effects by modulating the release of the following neuropeptide chemicals: acetylcholine, norepinephrine, dopamine, serotonin, and substance P. Besides relieving pain, opiates affect learning and memory, immune system function, appetite, gastrointestinal function and other autonomic nervous system functions such as respiration, temperature, cardiovascular function, and endocrine responses. In sufficiently high doses, opiates cause a generalized central nervous system depression capable of producing surgical anesthesia.

Opiate Agonists and Antagonists

Opiate (narcotic) agonists include the natural opium alkaloids such as morphine, codeine and heroin, semi-synthetic analog compounds such as hydromorphone, oxymorphone, and oxycodone, and synthetic compounds such as meperidine (Demerol), levorphanol, methadone, sufentanil, alfentanil, fentanyl, remifentanil, and levomethadyl. Also included are the natural or endogenous opiate agonists, the endorphins and enkephalins.

Mixed agonist-antagonist drugs have agonist activity for some opiate receptor subtypes and antagonist activity for other subtypes. Partial agonist drugs only activate some of the opiate receptor subtypes. Examples of these drugs include butorphanol and buprenorphine hydrochloride (Buprenex).

Buprenex acts as a partial mu opiate receptor agonist, meaning that it has components for analgesia and pain relief. However, because Buprenex is also a full kappa opiate receptor antagonist, it can manage short-term as well as chronic pain along with opiate detoxification. During detoxification, Buprenex allows comfortable, painless withdrawal without the fatigue, sweats, tactile sensation complaints (tingling, skin crawling), aches, seizure risks or confused thought processes common during traditional detoxification procedures.

Pure narcotic antagonists include naltrexone and naloxone and similar drugs which have no opiate receptor agonist properties. Opiate antagonists block the opiate receptor and inhibit the pharmacological activity of any opiate agonists that may be present.

Opiate Antagonists vs. Opiates

Opiate antagonists differ from opiates only by the replacement of the N-methyl substitute characteristic of opiates with an N-allyl, N-cyyclopropyl-methyl or related group. Antagonists can readily reverse the pharmacological effects of opiates. When combined with opiates, opiate antagonists are thought to reduce the addicting properties of opiate analgesics.

Biochemical studies of receptor binding have shown that although opiate agonists and antagonists compete for the same opiate receptors, these compounds interact with the receptor in different ways. For instance physiological concentrations of sodium enhance the binding of certain opiate antagonists and reduce the binding of certain opiate agonists. These and similar studies suggest that there are distinct binding sites for opiate agonists and opiate antagonists on the same opiate receptors.

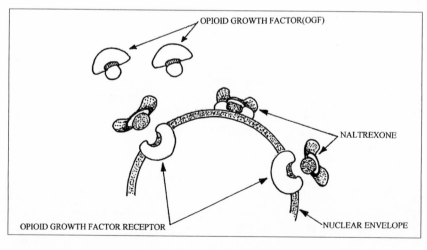

LDN blocking opiate receptors (Marvin G. Miller)

Ian Zagon's Early Studies

Since the early 1980s, Penn State professor Ian Zagon has been studying the relationship of opiates, endogenous opioid peptides, opiate antagonists, and the biochemical and cellular effects caused by these compounds. Doctor Zagon first became interested in the effects of opiate compounds when he observed reduced growth in infants born to mothers addicted to heroin. This led to Zagon's primary area of investigation, which involves studying the effect of opiate antagonists on cell growth, especially cancer cell growth and the cell growth involved in wound healing.

In the early 1980s, Zagon discovered that low dose naltrexone (LDN) increases the production of endogenous endorphins and increases the density of opiate receptors. Zagon's early publications on LDN inspired Bernard Bihari's studies and gave him a sound basis for his experimental use of LDN in human subjects.

Opioid Growth Factor (OGF) and Its Receptor (OGFr)

During the course of his studies, Zagon and his team discovered that the endogenous pentapeptide methionine enkephalin [Met5]-enkephalin also functions as a growth factor. Because of its functional properties, Zagon named this compound *opioid growth factor* (OGF). Pentapeptide endorphins (including Met-enkephalins, Leu-enkephalins, and proenkephalins), which are found in many parts of the brain, bind to specific receptor sites, some of which may be pain-related opiate receptors with properties of endogenous neurotransmitters and non-addicting analgesics.

Zagon also identified the receptor activated by OGF. He subsequently named it the *opioid growth factor receptor* (OGFr) and later determined its genetic sequence and cloned it. Zagon's team discovered that the axis or complex formed by OGF and the OGFr functions to regulate cell growth of both normal and abnormal cells.

LDN increases production of OGF as well as the number and density of OGF receptors by intermittently blocking the opiate receptor. The resulting intermittent increase in the OGF-OGFr complex repairs tissue and helps the body heal itself or return to a homeostatic state. Dr. Zagon's research involves studying increases in this complex of OGF and its receptor (either invoked by administration of LDN or by administration of pure OGF) and evaluating how it affects neuronal development, tissue repair, and cancer cell inhibition. Zagon's studies of opioid growth factor are described in chapter 2, and his research on low dose naltrexone in cancer is described in chapter 5.

FDA Drug Approval

All new drugs need proof that they're safe and effective before they can be approved for marketing. Before a new drug can be approved for the treatment of specific conditions, a set of clinical tests must be performed. In the preclinical research stage the drug is synthesized and purified. Then animal studies are performed and institutional review boards assess these studies. If the recommendations are positive, an application to the FDA is initiated and clinical tests are initiated.

Clinical Trials

Phase 0 trials, which were added in 2006, are designed to expedite the development of promising therapeutic or imaging agents by establishing very early on whether the agent behaves in human subjects as was anticipated from preclinical studies. A subtherapeutic dose of the drug is given to fifteen to twenty subjects to gather preliminary data on the drug's pharmacological properties.

Phase 1 clinical trials are clinical studies in which the investigational drug is tested on twenty to eighty humans in a clinical setting to observe structure-reactivity relationships, mechanism of action, and side effects. If possible, effectiveness is also evaluated.

Phase 2 clinical trials typically involve several hundred patients; these tests are used to determine the efficacy of a drug to treat patients with specific diseases or conditions and to learn about common short-term side effects or risks.

Phase 3 clinical trials explore the effects and safety of the drug on several hundred to several thousand people. Scientists then extrapolate how these results would pertain to the general population. Institutional review boards and advisory committees protect the rights of the clinical trial participants. The Center for Drug Evaluation and Research (CDER) uses experts to obtain outside opinions and advice and to provide new information on previously approved drugs.

Post-Approval Trials

Phase 4 trials involve the continued monitoring and surveillance of patients taking the drug once it has been released to the general public. Phase 4 trials include monitoring any reported adverse effects of a large number of patients who have been prescribed the drug.

Off-label Drug Use

Off-label or "unlabeled" drug use refers to the use of an FDA-approved drug for other uses than those listed in the FDA-approved labeling or in treatment regimens or patient populations that are not included in approved labeling. Evidence of efficacy in other conditions is not required for the new use of old drugs before they can be prescribed off label in these other conditions. Pharmaceutical manufacturers, however, are prohibited from marketing the off-label use of their products.

However, when there is an intention to report to the FDA a new indication for use or to support any other significant change in the labeling of a drug, an Investigational New Drug Application (IND) must be filed.[13]

Prescribing LDN

Naltrexone is an inexpensive, generic drug available in 50-mg capsules. No one pharmaceutical company can claim exclusive rights over it. The costs of conducting clinical trials and thereby receiving FDA approval for naltrexone's use in other conditions besides narcotic and alcohol abuse are prohibitive for pharmaceutical companies. Pharmaceutical companies are also prohibited from advertising compounds for uses that aren't FDA approved. Despite numerous anecdotal reports, as well as a modest number of clinical trials and published papers describing the efficacy of LDN in MS, AIDS, autism, cancer, neurodegenerative, and autoimmune diseases, funding for future trials of LDN remains limited. The largest trial for LDN in MS in the United States was partially funded by fundraisers organized by patients.[14]

Without clinical trials to prove efficacy in these off-label conditions, LDN continues to be prescribed off label. For the low doses used in LDN therapies, the drug must be obtained as a 50-mg capsule and diluted in water and divided into aliquots or it must be compounded by a compounding pharmacist and reformulated into the prescribed low doses, usually 3–10 mg. Unless the compounded preparation comes in a liquid form, various fillers are added to the drug suspension. Avicel (acidophilus and calcium carbonate), sucrose, lactose and other fillers are used, although some patients are intolerant of certain fillers, and some patients claim to respond better to certain fillers. (See chapter 9 for more information on LDN prescriptions.)

Many physicians are unfamiliar with LDN and some physicians are hesitant to prescribe LDN for the autoimmune, neurological conditions and cancers discussed in this book. In many medical circles, LDN is considered an alternative medical treatment. The interest in LDN primarily stems from patient accounts on the Internet. In Scotland, patients have petitioned their government for clinical trials for MS and demonstrated how much more cost effective LDN is than the conventional MS drug cocktail. Sadly, the potential uses of low dose naltrexone remain essentially unexplored despite the millions of dollars spent on clinical trials intended to improve naltrexone's position as an effective treatment for drug abuse and the millions of dollars spent on ineffective MS treatments.

Future Low Dose
Naltrexone Endeavors

Worldwide research into LDN continues. Within the last decade, several moderate-sized studies using LDN alone or with other compounds have been conducted, and more studies and clinical trials are underway. The handful of researchers whose interest in naltrexone was first sparked in the 1980s has grown.

In April 2007 the National Cancer Institute (NCI), a division of the National Institutes of Health (NIH), hosted a conference on Promising Compounds for Unmet Medical Needs that focused on opiate antagonists and endorphins. The NCI conference and also the annual LDN conferences that started in 2005 have given researchers worldwide an opportunity to present and share their findings.

Jill Smith's successful pilot trial of LDN in Crohn's disease was published in the online edition of the *American Journal of Gastroenterology* in January 2007 and in the print edition in April 2007. Burton Berkson's successful report of using LDN along with alpha lipoic acid long-term in one patient with pancreatic cancer was published in *Integrative Cancer Therapy* in March 2006. Ian Zagon has conducted dozens of trials involving the use of LDN and OGF in various cancers and diabetes, including pancreatic cancer, head and neck squamous cell cancers, colon cancer and renal cancer. The Mali LDN Trial for HIV/AIDS began in December 2007 and is expected to make a huge impact on the global fight against HIV and AIDS.

The following chapters describe the use of low dose naltrexone in specific conditions and explain the effects of LDN on immune system health and homeostasis.

2

LDN in Autoimmune Diseases

The widespread use of low dose naltrexone (LDN) can be largely attributed to anecdotal accounts published on Internet Web sites, chat groups, and bulletin boards describing its use in various autoimmune and neurodegenerative disorders. In particular, the success of LDN in multiple sclerosis has been widely publicized. Across the globe, a legion of Web sites created by both physicians and patients describe clinical trials and protocols for the use of LDN in MS and detail successful accounts of its use.

In print, publications as diverse as the *Townsend Newsletter for Doctors,* the *Accelerated Cure Project for MS, Medical Hypotheses,* the *European Journal of Neurology,* the *Annals of Pharmacotherapy,* and the *London Herald* have all printed articles describing the benefits of LDN in a variety of autoimmune diseases that emphasize the need for clinical trials.

However, many factors, including body weight, dose, disease severity, and metabolism, influence the potential benefits of LDN for individuals with autoimmune and other diseases. While anecdotal reports spark interest in patients and researchers alike, scientific research is the true measure. The results of clinical trials of LDN in humans along with animal studies demonstrating how LDN works on the body's tissues and cells are the hallmarks for determining its safety and efficacy.

Naltrexone and other opiate antagonists have been studied for their effects on human health long before the acronym LDN came into use. However, interest in LDN has skyrocketed in recent years primarily because of patients sharing their experiences with LDN on the Internet. Patient accounts of the use of LDN in nearly every autoimmune disease known abound on Internet bulletin boards and Web sites. And the number of clinical trials involving LDN in autoimmune diseases, while still small, has grown.

In an attempt to bridge the gap between anecdotal information and the clinical trials and research studies conducted on LDN, this chapter focuses on the nature of autoimmune diseases and describes the use of LDN in autoimmune conditions other than MS. The use of LDN in MS is the subject of chapter 3.

What Are Autoimmune Diseases?

Prior to 1957, autoimmune diseases were referred to as conditions of "horror autotoxicus," a term first used by the German physician Paul Ehrlich in the early 1900s. This term referred to the fact that in a number of disorders such as systemic lupus erythematosus (SLE) the immune system appeared to target "self" or "autologous" rather than "foreign" antigens. In other words the immune system appeared to be attacking its human host.

In 1957, the autoimmune hypothyroid disorder Hashimoto's thyroiditis was recognized as the first organ-specific autoimmune disorder. In the early 1960s the scientific study of the immune system, known as immunology, emerged. Experts in the field, immunologists, began studying the ways in which immune system dysfunction contributed to various types of disease. By 1965, autoimmunity began to be regarded as an important cause of human disease.

Autoimmune diseases include nearly one hundred different unrelated disorders that share a common link. In autoimmune disorders, the immune system launches an immune response that targets the body's own tissues and cells. Autoimmune diseases may be organ-specific, targeting specific organs such as the thyroid gland in the autoimmune hyperthyroid disorder Graves' disease; or they may be systemic, targeting multiple organs and bodily systems. Systemic lupus erythematosus (SLE) is an example of a systemic autoimmune disease. In SLE, the immune system may target one or more of the following organs: the lungs, joints, heart, kidneys, skin, and vascular system.

The Autoimmune Disorders

The organ-specific disorder known as autoimmune thyroid disease accounts for about 50 percent of all autoimmune diseases. Autoimmune thyroid diseases are primarily represented by Hashimoto's thyroiditis but also include Graves' disease, autoimmune atrophic thyroiditis, thyroid eye disease, Hashitoxicosis (condition of Hashimoto's thyroiditis in conjunction with the stimulating TSH receptor antibodies seen in Graves' disease, causing transient symptoms of hyperthyroidism), and Hashimoto's encephalopathy.

An additional 20 percent of all autoimmune disorders affect the joints, primarily in the form of rheumatoid arthritis and related rheumatologic conditions such as ankylosing spondylitis and psoriatic arthritis. Another 20 percent of all cases involve the kidneys (glomerulonephritis, Wegener's granulomatosis), the liver (autoimmune hepatitis, primary biliary sclerosis, sclerosing cholangitis), the red blood cells (hemolytic anemia), the salivary/lacrimal glands (Sjögren's syndrome), and the skin and eyes (scleroderma, uveitis).

The autoimmune disorders myasthenia gravis (affecting neuromuscular transmission), multiple sclerosis (affecting central nervous system transmission), insulin dependent diabetes mellitus (affecting insulin-producing cells of the pancreas), and SLE collectively account for only about 5 percent of the estimated nine million people in the United States with autoimmune disorders.[1]

The other 5 percent of autoimmune disorders are represented by a number of rare autoimmune conditions including Behçet's disease, pemphigus disorders, polyneuropathies, autoimmune polyglandular syndromes, polymyalgia rheumatica, disorders of myositis, adrenal insufficiency, Reiter's disease, Guillain-Barré Syndrome, vasculitis disorders, antiphospholipid syndrome, herpes gestationis, and celiac disease.

Who Develops Autoimmune Diseases?

Autoimmune diseases are reported to develop in individuals with certain predisposing immune and organ-specific genes when they're exposed to certain environmental triggers. Approximately 20 percent of the population is susceptible to autoimmune disease development. However, only 3 percent to 5 percent of the population goes on to develop autoimmune disorders.

Environmental Triggers

Stress and vitamin D deficiency have been reported to trigger and exacerbate symptoms in many of the autoimmune disorders. Other environmental triggers are associated with specific autoimmune disorders. For instance, excess dietary iodine and selenium deficiency are both associated with autoimmune thyroid disorders, and silica exposure is associated with scleroderma.

Other environmental triggers of autoimmune diseases include viruses, bacteria, heavy metals (such as cadmium, lead, and mercury), estrogens, silver, gold, silicone, mineral oil, airborne particulate matter (pollution), pristane, ionizing radiation, ozone, trauma, physical injury, pesticides, organophosphates, aspartame, dioxin, polychlorinated bromides, testosterone, cortisone, diethylstilbeserol (DES), pentamidine, L-asparginase, pyriminil, vaccines, canavanine, and halogenated hydrocarbons.[2]

Gender Influences

Overall, females are about seven to nine times more likely than men to develop autoimmune disorders, and 75 percent of all autoimmune disorders occur in women. However, gender influences vary in different autoimmune disorders. For instance, while 90 percent of autoimmune thyroid diseases occur in women, in the rheumatologic condition ankylosing spondylitis most disorders occur in men.

Age Influences

Although people of all ages can be affected by autoimmune disorders, young to middle-aged adults are the primary targets. The risk of developing autoimmune diseases also increases with age as immune function begins to falter. Again, this varies with specific disorders. For instance, ankylosing spondylitis tends to occur in adolescent and young adult males, and it's known to develop in males as young as ten. In contrast, the autoimmune rheumatologic disorder polymyalgia rheumatica nearly always develops in people older than fifty.

Ethnic Differences

Autoimmune diseases are primarily seen in the western world. Immigrants to the West, however, take on the same incidence of the disease as the native or indigenous population, often exceeding it. Whether the differences in autoimmune disease incidence between western and Third World nations are due to sociological, nutritional, parasitic or hygienic reasons remains unclear. According to researchers the primary candidates for these differences include infectious agents, vaccines, chemicals, diet, and other environmental factors.

Specific autoimmune diseases also have an increased incidence in certain populations. For instance, in the United States, ankylosing spondylitis has the highest incidence in American Indians. And although multiple sclerosis occurs worldwide, it is most common in Caucasian people of northern European origin, especially those of Scottish descent. MS is extremely rare among Asians, Africans, and Native Americans. In specific groups (Rom, Inuit, Bantus) no cases have ever been reported. While the risk of MS for African Americans is around half of that for Caucasian Americans, a recent study at the University of California, San Francisco, suggested that African Americans are more likely to develop a more aggressive form of MS and to suffer impaired mobility.[3]

Theories of Autoimmune Disease

Various theories have been proposed to explain why the immune system attacks itself in autoimmune disease. Most experts, however, agree that autoimmune diseases are clearly heterogeneous disorders. That is, different autoimmune disorders are caused by different factors, so, no one theory explains why all autoimmune disorders develop, although various abnormalities in immune components such as cytokines, autoantibodies, immune complexes, and white blood cells all contribute to autoimmune diseases in specific disorders.

In fact, some researchers have questioned whether all disorders classified as autoimmune truly have an autoimmune origin. In particular, the question of whether Crohn's disease and multiple sclerosis are autoimmune disorders has been debated in several publications.[4] While a discussion of this debate is outside the scope of this book, a description of autoimmune disease theories is useful from an academic standpoint, and helpful for understanding how LDN's effects stop disease progression in a variety of different disorders mediated by defects in immune function.

Hyperactive Immune Function

Older theories of autoimmune disease proposed that the immune system was hyperactive, strong, and easily stimulated. Today, it is generally accepted that the immune system in autoimmune disease is weak and ineffective. And because many healthy people show evidence of autoimmunity, it's suspected that in autoimmune disease the body's innate ability to keep autoimmunity in control is defective or inhibited.

Weakened Immune Function

In some autoimmune diseases, such as Crohn's disease, Graves' disease, and type 1 diabetes, the immune system is thought to be weakened by chronic exposure to infectious agents (*Yersinia* and *Campylobacter* in some patients with Graves' disease and viral particles in patients with insulin-dependent diabetes mellitus); vaccines (influenza vaccine in patients with Guillain-Barré syndrome); chronic exposure to chemicals (Wegener's granulomatosis) and chronic exposure to allergens, antibiotics, and airborne pollutants.

It's thought that a weakened immune system cannot launch an appropriate immune response. Instead, the immune response is erratic and ineffective, which leads to the continuous targeting of self rather than foreign antigens.

Oxidative Stress and Autoimmunity

The immune system relies on unstable molecules known as free radicals to destroy invading organisms and malignant cells. Free radicals are molecules that lack an electron in their outer ring. To stabilize themselves, they steal electrons from other molecules, causing the production of more free radicals. Free radicals are produced by the environment or they may be human-made. For instance, free radicals may be produced by exposure to cosmic radiation, cigarette smoke, or the body's own metabolic processes. Free radicals include hydroxyl, superoxide, and nitric oxide radicals. Other molecules known as reactive oxygen species, including hydrogen peroxide and peroxynitrites contribute to oxidative stress by enabling free-radical production.

In autoimmune disorders the chronic activation of the immune system leads to increased free radical activity and an increase in oxidative stress, an

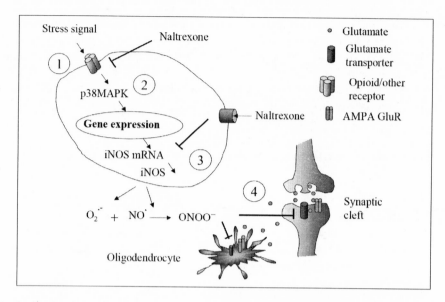

Reduction in oxidative damage. (Reprinted from *Medical Hypotheses*, 64, Yash P. Agrawal, M.D., "Low Dose Naltrexone Therapy in Multiple Sclerosis," p. 4, 2008, with permission from Elsevier.)

increase that cannot be countered by the body's own antioxidant enzymes. This chain of events is thought to contribute to the development and perpetuation of autoimmune diseases.

Apoptosis

All of our body's cells are programmed to die at a specified time in a process of programmed cell death known as apoptosis. Immune-mediated changes in apoptosis of the brain's cells (neurons) are thought to contribute to the neuronal damage seen in neurodegenerative diseases such as MS, Parkinson's disease, and Alzheimer's disease. According to this theory, microglia (immune system cells of the central nervous system) are believed to assist in the clearance of cells programmed for elimination through apoptosis. However, microglia become readily activated in response to infection or injury.

During this activation, microglia secrete a variety of soluble pro-inflammatory and neurotoxic factors. Excessive quantities of these factors related to excessive microglial stimulation contribute to neurodegeneration and the development of lesions in MS. Lesions in MS show apoptosis of oligodendrocytes (cells that produce myelin) and activation of microglial cells. Considerable evidence suggests that the oligodendrocyte cell apoptosis, demyelination, and axonal damage seen in MS lesions may be related to oxidative stress and neural toxicity associated with excitotoxins such as glutamate that stimulate microglial activation.

Excitotoxins

Excitotoxins are neurotransmitters that stimulate neurons and also the oligoendrocyte cells that produce the myelin sheath covering nerve fibers. By stimulating neurons, excitotoxins aid in cognition, sight, taste, and hearing. The primary excitotoxin found in the brain is glutamic acid or glutamate. Glutamate stimulates the brain by allowing more calcium to enter neurons.

In high concentrations, glutamate accumulates, for instance when there are defects in the blood-brain barrier caused by injury or stroke, defects in the glutamate transporter, or metabolic changes such as hypoglycemia. This excess quantity of glutamate damages neurons and oligodendrocytes. Excitotoxins are reported to cause neurodegenerative changes that contribute to the development of multiple sclerosis, Parkinson's disease, and Alzheimer's disease.[5] Low dose naltrexone causes changes that prevent glutamate from accumulating. The role of low dose naltrexone in reducing concentrations of glutamate and excitotoxins is described further in the following section and in chapter 4 with neurodegenerative diseases.

LDN and Glutamate Reduction

The Cornell University School of Medicine professor Dr. Yash Agrawal reports that direct evidence of glutamate toxicity may be seen in multiple sclerosis. Agrawal writes that the developing lesions in MS do not show evidence of inflammation or the presence of inflammatory cells. Rather, the damage in MS may reflect oxidative stress mediated by nitric oxide synthase, nitric oxide, and peroxynitrites produced by activated microglial cells. Microglial cells are the resident immune cells of the central nervous system.

Low dose naltrexone, by reducing nitric oxide synthase activity, decreases the formation of peroxynitrites, which, in turn, prevents the inhibition of glutamate transporters, reducing glutamate formation and accumulation, thereby reducing excitatory neurotoxicity, and oligodendrocyte apoptosis.[6]

In a related report, Dr. Jau-Shyong Hong, Director of the Neuropharmacology Department of the National Institute of Environmental Health Sciences (NIEHS) division of the National Institutes of Health (NIH), writes that "glutamate is a potent and rapidly acting neurotoxin on cultured spinal neurons."[7] In studies of mixed spinal cord neuron-glia cultures, cells exposed to glutamate for five minutes produced neuronal swelling, followed by neurodegeneration (cell damage and destruction) within twenty-four hours. Hong is also one of the country's leading experts on the use of opiate antagonists in neurodegenerative disorders.

Infectious Agents and Molecular Mimicry

The autoimmune disease theory known as molecular mimicry is based on the premise that viruses and bacteria have the ability to mutate and change their physical characteristics. In a process of molecular mimicry designed to enhance their survival, these infectious microorganisms can take on characteristics that mimic those of their human host. Consequently, an ineffective immune system, which still perceives the presence of foreign antigens in the mutated molecules, is unable to distinguish between the infectious and self molecules and begins to attack the body's own cells.

Dr. Noel Rose, the country's leading expert on autoimmune disease, joined other researchers for a colloquium on cell damage and autoimmunity in April 2007 at the Johns Hopkins Center for Autoimmune Disease Research. Rose described the major causes of autoimmune disease as inflammation, infection, apoptosis, environmental exposure and genetics. Rose reported that autoimmune diseases documented as occurring as sequelae to infection include rheumatic fever and Guillain-Barré type polyneuropathy, and possibly late-phase Lyme disease. Suspect infections requiring closer scrutiny for their role in

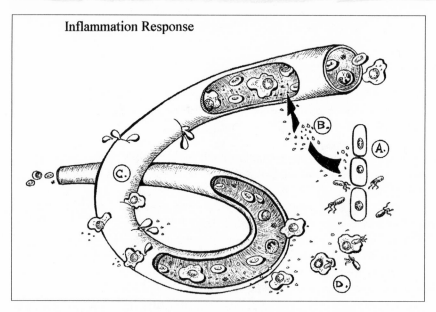

Inflammation response (Marvin G. Miller). (A) Bacteria or other irritants invade and damage tissue. (B) Chemicals, such as histamines and cytokines, are released by tissue cells and enter blood vessels. (C) The blood vessel wall becomes more elastic and permeable, allowing fluid and activated macrocytes to enter the affected area. (D) Phagoctyes cells attack the bacteria; serum complement and other protein molecules enter the affected area. (E) Secondary inflammatory reaction follows, including heat, swelling and pain.

autoimmune disease development include the Epstein Barr virus (systemic lupus erythematosus and other diseases), hepatitis C virus (autoimmune hepatitis), beta hemolytic streptococci (rheumatic carditis), and *Heliobacter pylori* (autoimmune gastritis).[8]

Cytokine Influences

Cytokines are immune-system chemical messengers that modulate the immune response. Cytokines include interferon, interleukin, and other immune-system chemicals, which are released during the immune response. Certain cytokines promote inflammation and facilitate the activation of lymphocytes that react with the body's proteins.

Cytokines released during the allergic response can trigger and exacerbate symptoms in patients predisposed to autoimmune disease development. For instance, in patients with autoimmune thyroid disease who have gluten (protein in wheat, rye, and barley) sensitivity, levels of both gliadin (related to gluten sensitivity) and thyroid antibodies increase when gluten is ingested. Allergens are also a common trigger for autoimmune diseases. Allergies are considered the primary cause of autoimmune asthma. In Japan, allergies to cedar pollen are considered the primary environmental trigger leading to Graves' disease.

Neuroimmune Influences

The immune system is constantly interacting with the neuroendocrine system. These bidirectional interactions influence antibody and cytokine responses, cytolytic activity, lymphocyte proliferation, tissue localization of lymphocytes, hypothalamic-pituitary hormone secretion, and neural-signal transmission.[9] This is best exemplified in the body's response to stress. Depending on the duration and type of stressor, stress can either suppress or enhance immune responses.

For instance, numerous studies have shown that exhausting physical activity and severe environmental and/or psychological stress have strong suppressive immune system effects with significant implications for disease susceptibility and progression. Investigations in both humans and animals show that chronic stress could promote tumor development, autoimmunity, and infectious diseases by influencing the onset, disease course, and outcome of the disease pathologies. As an example, as many as 77 percent of patients with Graves' disease have experienced stressful events before disease onset, and one of the early symptoms in Graves' disease is increased nervousness, which appears to have been triggered by events such as surgery, pregnancy, traumatic experiences or bereavement.[10]

Acute psychological stressors and moderate physical stress, on the other hand, transiently enhance immune responses. Acute stress can also increase antibody production and enhance delayed-type hypersensitivity reactions. By increasing endorphin production and enhancing well-being, naltrexone is thought to improve the effects associated with chronic stress.

LDN in Crohn's Disease

Crohn's disease is a severe autoimmune inflammatory disease of the small bowel that may involve any part of the intestine, including the colon. Often, the entire bowel wall is involved. Unlike ulcerative colitis, some parts of the intestine are affected while adjacent areas remain unscathed. Lesions in Crohn's disease involve the presence of strictures, abscesses, and fistulas. In 2004, Moshe Rogosnitzky, director of research at MedInsight Research Institute, presented evidence to Ian Zagon at Penn State suggesting that increased endorphin levels could benefit patients with Crohn's disease.[11]

As a result, in 2006, Jill Smith and colleagues at Penn State conducted the first open-label placebo-controlled pilot study of LDN in Crohn's disease. Eligible patients with active Crohn's disease and Crohn's Disease Activity Index (CDAI) scores equal to or greater than 220 were allowed to continue on anti-inflammatory medications but were prohibited from taking the tumor necrosis factor alpha (TNF-α) drug infliximab (Remicade).

At the study's end, two-thirds of the patients in the pilot study went into remission and 89 percent of patients reported improvement to some degree based on the CDAI score, quality of life surveys, and blood tests for inflammatory markers. The results of the pilot trial, published in the January 2007 online edition and the April 2007 printed edition of the *American Journal of Gastroenterology*, emphasized the role that endogenous opiates and opiate antagonists such as LDN play in the healing and repair of tissues. Of the seventeen patients given LDN who completed the study, two subjects with open fistulas experienced fistula closures. Two subjects discontinued routine medications; of these one experienced a flare-up of symptoms and the other subject was maintained on LDN alone.

Overall, after twelve weeks of LDN, followed by four weeks off medication, there was a significant improvement in CDAI and Quality of Life scores with minimal side effects. The most common side effect reported was sleep disturbances, which were reported in seven of the seventeen patients enrolled in the study.[12]

As a result of her success in the pilot trial, Smith received funding from the National Institute of Health (NIH) and the Broad Medical Research Foundation to pursue a phase 2 trial of LDN in Crohn's disease. In February, 2008 Smith opened registration and began recruiting participants for the first phase 2 clinical trial of LDN at a U.S. medical center. The results of the phase 2 trial are expected to confirm the potency and safety of LDN in the treatment of autoimmune disease.

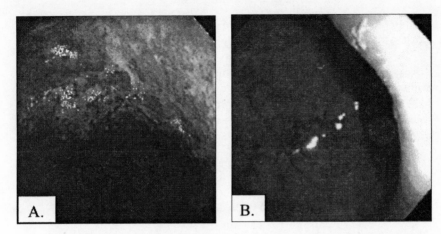

Crohn's disease, clinical trial results (Jill Smith, M.D., Pennsylvania State University).

On his Web site, Bernard Bihari reports having treated eight patients with Crohn's disease with LDN as of September 2002. All patients experienced improvement within fourteen to twenty-one days of starting treatment and have remained in remission from five to eight years since starting treatment.

Related Animal Studies of LDN

In a related ongoing study at Penn State, Smith and her colleagues are studying the chemical and molecular mechanisms involved in suppression of inflammatory responses in the intestine of animals treated with naltrexone. Dextran sulfate sodium (DSS) was administered to mice in their drinking water to induce colitis. The mice were then treated with naltrexone for six days after which time their colons were excised and tissue studies were performed. The RNA was extracted from the cells and evaluated for cytokine expression. The investigation showed that naltrexone reduced colonic inflammation in the animals.[13]

Inflammatory Bowel Disease (IBD) and Ulcerative Colitis

Inflammatory bowel disease (IBD) is a collective term that refers to chronic, autoimmune, inflammatory diseases of the bowel, mainly ulcerative colitis and Crohn's disease, although IBD may also be referred to as colitis, enteritis, ileitis, and proctitis.

Patients with IBD have symptoms of diarrhea, fever, abdominal pain, usually on the right side of the lower abdomen, feeling of a mass or fullness in the lower, right abdomen, weight loss, and bloody stools. Perinuclear antineutrophil cytoplasmic antibodies (P-ANCA) and inflammatory white blood cells are seen in the colon mucosa of patients with IBD. On occasion, signs and symptoms of both ulcerative colitis and Crohn's disease both occur in patients, causing an overlapping type of inflammatory bowel disease.

Ulcerative colitis primarily involves the mucosal lining of the colon. The disease typically affects the rectum and spreads to the proximal colon. The entire colon is affected in some people. In others, disease is restricted to the lower areas of the bowel. Like most autoimmune disorders, periods of symptoms tend to alternate with periods of remission. Ulcerative colitis is associated with an increased risk of colon cancer particularly in patients with extensive colitis of long duration.

Irritable Bowel Syndrome (IBS)

Although it is not an autoimmune disorder, irritable bowel syndrome (IBS) often occurs in people with autoimmune diseases and it is has been shown to respond well to treatment with LDN. IBS is a common functional gastrointestinal disorder causing increased contractions or spasms of the colon or rectum. In people with IBS, the colon is more sensitive and contracts more readily than it does in other people. IBS is a syndrome or collection of symptoms rather than a disease.

It is also a condition of spastic colon rather than an inflammatory disorder. Symptoms of IBS include chronic abdominal pain, discomfort, diarrhea, and constipation, and they often occur in people with systemic lupus erythematosus (SLE), fibromyalgia, and other disorders. In these conditions, symptoms of IBS are often linked to stress, inadequate fluids, poor nutrition, and poor eating habits, such as eating too fast or not relaxing after a meal. No tissue changes are seen in the colon of patients with IBS.

IBS is diagnosed in people having symptoms for at least twelve weeks within the preceding twelve months. Two of the following features must be present to diagnose IBS: relief upon defecation, onset associated with a change in stool frequency, and onset associated with a change in stool form. Imaging and blood tests are used to rule out the presence of other disorders that may mimic IBS.

Incidence of Inflammatory Bowel Disease and Irritable Bowel Syndrome

Both IBD and IBS primarily emerge in young people between ten and twenty or people in their early thirties. Up to 600,000 Americans (1.5 percent of the population) are reported to have irritable bowel disease, with two-thirds of all cases occurring in women.[14]

Coexistence of IBS and IBD

Although the topic is still controversial, many physicians think that symptoms of IBS can occur in patients with IBD. It's also known that patients with IBD often complain of having symptoms of IBS for several years before being diagnosed with IBD. Studies show that people with IBS do not have a higher risk of developing IBD than other people. However, people diagnosed later in life with IBD tend to have a history of IBS symptoms prior to diagnosis.

LDN in IBS

In a controlled open-label trial of LDN in patients with IBS, oral doses of 0.5 mg naltrexone were administered to forty-two patients daily at bedtime for four weeks, followed by a four-week follow-up period. Study participants received naltrexone in the form of PTI-901, a compound manufactured by the biopharmaceutical company Pain Therapeutics, Incorporated. During treatment, patients were examined weekly and their self-reported assessment scores were reviewed. Patients were not allowed to take opiates up to thirty days before the trial. Antispasmodic and antidiarrheal agents were allowed during the trial.

A total of thirty-six patients completed the study and global response to treatment was 78.5 percent for men and 72 percent for women. Improvement was seen in more than 83 percent of all patients, which included patients with diarrhea-predominant IBS, constipation-predominant IBS and alternating IBS. The mean pain-free days per week significantly increased in all patients. In this trial LDN was well tolerated. The data suggest that LDN can block excitatory opioid receptors without affecting inhibitory opioid receptors, offering analgesic potency and pain relief without side effects.[15]

In a follow-up open-label Phase 2 study, fifty patients were treated with LDN in a four-week study. Patients were evaluated by self-assessing their improvement using the Global Assessment of Adequacy of Treatment questionnaire. Overall, patients reported a 140 percent increase in the number of pain-free days they experienced at week four compared to baseline. Clinical improvements in bowel urgency, stool consistency and number of stools per day were also reported at week four in both genders, and PTI-901 was reported to be well tolerated.[16]

Rheumatoid Arthritis and Other Connective Tissue Disorders

Dr. Bihari reports that ten of his patients with rheumatoid arthritis have shown a good response to LDN. In all patients, joint pain and swelling were reduced although some patients still had residual joint distortion. Two patients who stopped LDN for several weeks reported immediate exacerbations. Another patient who showed a good response noticed a mild exacerbation of symptoms during a period of severe marital stress.

Bihari also reports using LDN to halt disease progression in patients with ankylosing spondylitis and systemic lupus erythematosus. The rash of psoriasis and conditions of eczema are also reported to improve with LDN. Topically, used in high doses (HDN), naltrexone aids wound healing, a topic described in chapter 7.

Fibromyalgia

Fibromyalgia is a musculoskeletal disorder causing widespread pain in muscles, tendons, and ligaments, and pain upon pressure of specific trigger points. Headaches, sleep disturbances, and symptoms of irritable bowel syndrome (IBS) are also common. Fibromyalgia was first described in 1990 by Dr. Frederick Wolfe, the director of the National Databank for Rheumatic Diseases, in a 1990 paper that first defined fibromyalgia's diagnostic guidelines.

The causes of fibromyalgia are unknown although it is suspected that patients with fibromyalgia have a threshold for tolerating pain impulses that is substantially lower than that of most individuals. This lowered pain threshold is suspected of being related to sensitivity of brain cells to pain signals.

LDN has been reported anecdotally to reduce the symptoms of fibromyalgia presumably by regulating natural pain-reducing symptoms. Jarred Younger, a postdoctoral fellow in the School of Translational Medicine, and his colleagues at Stanford University School of Medicine, are investigating this premise in a double-blind, placebo-controlled clinical trial involving patients with moderate to severe fibromyalgia, aged eighteen to sixty-five.

Younger theorizes that in fibromyalgia LDN works by reducing the activity of microglia, the resident immune system cells of the central nervous system, by blocking opiate receptors on these cells. In doing so, the persistent microglial activation that is suspected of occurring in fibromyalgia is reduced. The reduced microglial activation in turn reduces the high levels of circulating pro-inflammatory cytokines seen in patients with fibromyalgia.

Patients in the ongoing trial are being evaluated for pain, fatigue, sleep quality, mechanical pain sensitivity, and thermal pain sensitivity. The trial is being funded by the American Fibromyalgia Syndrome Association as well as private gifts. The results of Younger's trial are expected to be available by late 2008.[17]

Chronic Fatigue Syndrome

Chronic fatigue syndrome (CFS) is an autoimmune disorder character-ized by muscle pain and sleep disturbances. Low dose naltrexone has been found to reduce pain and improve sleep in patients with CFS. One patient with CFS who participated in an informal eight-week trial using 3 mg of LDN at night reported significant improvement.[18]

Bihari's Autoimmune Disease Studies

On his Web site, Dr. Bihari reports that nearly twenty years ago, while conducting his LDN trial in patients with HIV/AIDS, his daughter's close friend suffered three acute episodes of multiple sclerosis within a three-month period. Because of the autoimmune component of MS and his experience with LDN, he started his daughter's friend on 3 mg naltrexone taken orally at bedtime. The girl's symptoms improved and disease progression stopped. When the girl quit taking LDN, her symptoms worsened, so she resumed LDN therapy, which she continues to use to prevent disease progression.

Since the late 1980s Bihari has treated numerous patients with autoim-mune disorders, including chronic fatigue syndrome, Crohn's disease, fibro-myalgia, multiple sclerosis, psoriasis, rheumatoid arthritis, pemphigus, autoimmune thyroid disease, amyotrophic lateral sclerosis, lymphoma, scle-roderma, systemic lupus erythematosus, ulcerative colitis, celiac disease, emphysema, chronic obstructive pulmonary disease (COPD), endometriosis, irritable bowel syndrome, sarcoidosis, transverse myelitis, and Wegener's gran-ulomatosis, using doses ranging from 3 mg to 4.5 mg taken daily at night. Bihari describes LDN as stopping disease progression rather than curing dis-ease and reports that most patients show improvement either using LDN alone or in conjunction with other remedies.[19]

Bihari has received patents for the use of low dose naltrexone in multi-ple sclerosis, prostate cancer, lymphoproliferative syndrome, herpes, human immunodeficiency syndrome, and chronic fatigue syndrome.

Additional Uses of LDN

Across the globe, physicians are using LDN for MS and other autoimmune disorders. Some of these physicians are listed in chapter 9. In this section we list several other researchers who are known for using LDN in autoimmune diseases.

Dr. Burton Berkson of Las Cruces, New Mexico, has presented several papers at annual LDN conferences describing his successful use of LDN in patients with autoimmune diseases. Doctors Fred Sherman and David Atkinson have received patents for their use of naltrexone in autoimmune disease, including systemic lupus erythematosus and rheumatoid arthritis. A description of their treatment protocol can be found at Free Patents Online.[20]

On RevolutionHealth.com, Dr. Trent Nichols and his colleagues have posted a large collection of anecdotal patient reports on LDN. Here, Sandi Lanford of the Lanford Foundation, an educational Lyme disease group, reports a wide range of autoimmune disorders for which LDN has been used.[21]

Because of the common genetic link seen in autoimmune thyroid disorders and rheumatoid arthritis, a number of physicians are using LDN to reduce inflammation and thyroid antibody titers in patients with autoimmune thyroid disease and thyroid eye disease.

Dr. Thomas Hilgers at Creighton University is using LDN experimentally as a treatment for autoimmune infertility disorders.[22] His theory is that many autoimmune diseases result from endorphin deficiencies. Endorphins are known to modulate the immune system. LDN is known to increase endorphin levels, thereby modulating the immune system.

3

LDN in Multiple Sclerosis

Patients with multiple sclerosis are true pioneers when it comes to LDN. Of all the groups that have embraced LDN, patients with MS have shown the most initiative in taking charge, initiating clinical trials, publishing newspaper articles, arranging radio broadcasts, starting informational Web sites, and approaching their doctors about its use. It's no wonder that many doctors who regularly prescribe LDN do so exclusively for patients with multiple sclerosis.

Patients with MS have also gone through extraordinary lengths to obtain LDN. For many patients, acquiring LDN has been their last hope for restoring their health and stopping disease progression. Mary Boyle Bradley's poignant story of procuring LDN for her husband is the focus of her book *Up the Creek with a Paddle.* Samantha Wilkinson's LDN story of starting LDN when her doctor refused to prescribe it for her progressive form of MS is described in chapter 9.

This chapter explains what distinguishes MS from other autoimmune and neurodegenerative conditions and it describes current theories about the development of MS. In addition it describes animal studies and clinical trials that support the use of LDN in treating MS.

What Is Multiple Sclerosis?

Multiple sclerosis, which was once called disseminated sclerosis or encephalomyelitis, is an autoimmune, inflammatory, demyelinating disease affecting the central nervous system. Demyelinating means that there is a loss of or damage to the protective myelin sheath that coats nerve fibers. The term multiple sclerosis, which means "many scleroses," refers to the multiple plaques of scar tissue (lesions) seen in the white matter of the central nervous system in patients with MS. The disease targets the neurons in the white matter of the brain and spinal cord, interfering with the signals sent by cells in between the gray matter and other organs in the body.

Pathology in MS

Specifically, MS destroys the oligodendrocyte cells that make up the myelin sheath, the fatty layer that helps neurons carry electrical signals. MS causes a thinning or complete loss of the myelin sheath. Sometimes it results in a cutting (transection) of the neuron's extensions, which are known as axons. Without myelin, neurons are unable to conduct their electrical signals effectively.

Loss of myelin in the lesions of MS accounts for some of the symptoms. For instance, damage along the complex nerve pathways that coordinate movement results in tremor. However, much of the damage seen in MS occurs outside of these regions. For instance, urinary tract infections can occur as a secondary symptom related to bladder incontinence.

Discovery of MS

Dr. Jean Martin Charcot (1825–1893) was the first person to scientifically describe, document, and name the disease process called multiple sclerosis. Prevalence[1] today is estimated at 400,000 in the United States, and over 2.5 million worldwide. MS symptoms can begin from ten to sixty years of age, but it usually starts between twenty and forty. Women outnumber men by a ratio of two to one.

Disease Course

The early stage of MS is characterized by recurrent bouts of inflammation, referred to as attacks, exacerbations or flares that target the central nervous system (CNS). A true exacerbation of MS is caused by an episode of inflammation in the central nervous system. Inflammation is followed by demyelination, which, in turn, results in the formation of an abnormal area called a plaque or lesion. Lesions slow, halt, or distort the nerve impulse. An example of an exacerbation of MS would be the development of optic neuritis, an inflammation of the optic nerve that impairs vision.

The clinical course of the disease varies depending on the subtype of disease present. MS also varies in disease severity, and symptoms are known to be exacerbated by heat, stress, upper respiratory infections, and flu-like illnesses. Similar to other autoimmune diseases, MS is associated with symptoms that wax and wane, often appearing during times of stress. This waxing and waning of symptoms can make diagnosis difficult.

Multiple Sclerosis Subtypes

MS is categorized into different phenotypes. Relapsing remitting MS (RRMS) is the most common form of MS and accounts for about 85 percent of cases. RRMS is characterized by clinical relapses that occur every one to two years. The definition of relapse requires that a new symptom or sign be present for at least twenty-four hours, and not be associated with a fever or intercurrent illness (such as influenza or a urinary tract infection), as elevated body temperature can unmask silent or old lesions. The severity in RRMS varies. Patients with severe MS may have more frequent flares.

Within the RRMS group, 50 percent of patients go on to develop the secondary progressive MS (SPMS) form of the disease within ten years of their initial diagnosis. SPMS follows a steadily worsening disease course with or without occasional flare-ups and minor remissions. This subtype involves less active inflammation. However, it's associated with significant chronic degenerative changes.

About 15 percent of patients with progressive MS have gradually worsening manifestations from the onset without clinical relapses and are described as having primary progressive MS (PPMS). Patients with PPMS tend to be older, and generally respond less effectively to standard MS therapies.

The progressive relapsing form of MS (PRMS) occurs in only about 5 percent of cases. This uncommon form is progressive from the onset with superimposed relapses (with or without recovery).

Diagnosis

Diagnosis and monitoring of MS are based on evoked potential results, imaging tests of the brain and spinal column by MRI, and cerebrospinal fluid (CSF) analysis via lumbar puncture. Up to 85 percent of patients with MS show evidence of abnormal proteins known as oligoclonal bands in their cerebrospinal fluid.

The degree of disability in MS is usually assessed with the expanded disability status scale (EDSS). The diagnostic tests for MS are also useful in determining the rate of disease progression and the effectiveness of therapies, including the evaluation of treatments being assessed in clinical trials.

Causes of MS

No definitive cause for MS has yet been identified. In fact, the host of candidates pursued by researchers is wide and varied. The Accelerated Cure Project for Multiple Sclerosis (ACP)[2] is a national nonprofit dedicated to curing MS by determining its causes. Its research path is determined by "the Cure Map," which systematically first sets up the casual candidates by the five known root determinants of any disease: genetics, pathogens, nutrition, toxins or trauma. Genetics and pathogens garner the most contenders for MS. ACP maintains exhaustive Cure Map Documents in the downloads section of their Web site that tracks completed and ongoing studies in each of these specific categories.

Infectious Agents

Many viruses, particularly the Epstein-Barr virus (EBV), human herpes virus-6 (HHV-6), and *Chlamydia pneumoniae* bacteria have been implicated as causes of MS. However, because evidence of past infection with these viruses is not uncommon, and most people with evidence of past infection with these organisms do not go on to develop MS, this matter is still under debate.

Patients with MS are also reported to frequently show evidence of chronic Lyme disease, a condition that is also sometimes misdiagnosed as MS. In his book, *The Lyme Disease Solution*, Ken Singleton explains how infection with *Borrelia*, the microorganism responsible for Lyme disease, can cause an immune dysregulation that causes susceptibility to autoimmune diseases.[3] Of interest, patients with MS and patients with chronic Lyme disease show favorable responses to LDN.

Patients with MS are suspected of having high levels of reactive oxygen species as a result of the pro-inflammatory immune response to chronic infection. Oligodendrocytes, the cells that produce myelin, seem to be particularly vulnerable to oxidative stress. There is some evidence that the progression of MS is more severe in certain patients who have a genetically determined inability to clear the products of oxidative stress. LDN's role in reducing oxidative stress is described in chapter 4.

Treatments

Drug therapy ranges from medications used to treat specific symptoms such as bladder dysfunction, pain and spasticity, to drugs that have been approved specifically for MS that were developed in the last decade and designed to slow attacks and reduce disease progression. No drugs have yet been developed that can reverse nervous system damage that has already occurred, and no therapy has been developed for the most serious form of MS, PPMS.

Since 1993, the U.S. Food and Drug Administration (FDA) has approved several drugs designed to slow disability and disease progression in patients with RRMS, including those with secondary progressive disease who continue to have relapses. Large multicenter placebo controlled trials rate their effectiveness at about 33 percent. Side effects range from flu-like symptoms to adverse reactions, including the development of other autoimmune diseases such as insulin dependent diabetes mellitus and autoimmune thyroid disease that preclude their use for some patients. These agents and their delivery methods include:

- IFNβ-1b (Betaseron) by alternate-day subcutaneous injection
- IFNβ-1a (Avonex) by weekly intramuscular injection
- IFNβ-1a (Rebif) by thrice-weekly subcutaneous injection
- glatiramer acetate (Copaxone) by daily subcutaneous injection

Immunosuppressive treatments approved by the FDA are reserved for more serious forms of SPMS, or worsening RRMS:

- mitoxantrone (Novantrone) by IV every three months for two years
- natalizumab (Tysabri) by IV every four weeks ongoing

Mitoxantrone is a chemotherapeutic agent with serious risks, including congestive heart failure[5] and leukemia.[6] Natalizumab, a monoclonal antibody, is the most recently approved drug, but it is only recommended for patients who fail all other disease-modifying therapies, due to the risk of experiencing reactivation of Jakob-Creutzfeldt disease, or developing a rare and fatal viral disease, progressive multifocal leukoencephalopathy (PML).[7] This drug was eagerly anticipated due to early reports of effectiveness and even given expedited approval. The studies were then suspended after PML occurred in three clinical trial participants. Natalizumab was rereleased in 2006, in restricted distribution.

Patient Dissatisfaction

The FDA-approved MS drugs don't make patients feel better since they do not typically improve symptoms. At best current MS therapies are reported to be 30 percent effective in improving symptoms and have multiple adverse effects. These drugs, which require injections, also carry a high price tag ranging from $15,000 to $25,000 per year, making them primarily available only to those with health insurance, or those who can get into an assistance program sponsored by the drug makers. Patricia Coyle, M.D., a leading MS researcher, estimates that 7–49 percent of patients proceed to try the entire collection of MS drugs before they're finally given the distinction of being designated as "non responders."[8]

The Attraction of LDN

Because effective treatments are lacking, LDN has caused a stir in the MS patient community. Who wouldn't be excited about an inexpensive generic drug with the potential to halt disease progression, and that has little to no side effects or risks? In addition, because it increases levels of endogenous opiates, LDN elevates mood. And although effects may not be immediately apparent, most patients with MS report noticing an improvement in symptoms. LDN is also reported to be effective for primary progressive MS (PPMS) a condition stubbornly resistant to most other MS therapies. A survey of more than 1300 MS patients at the *This is MS* Web site shows that LDN is more commonly used than Betaseron and is used about as often as Rebif.[9]

The possible mechanisms by which LDN could improve the course of MS are described in chapters 2 and 4. These effects include reduced oligodendrocyte destruction by apoptosis, reduced microglial activation, and anti-inflammatory and antioxidant effects. The early results of human trials suggest that LDN does offer efficacy in MS, and animal studies have begun to shed light on the biochemical effects that reduce demyelination and disease progression.

Dr. Bihari and MS

According to Internet legend, nearly twenty years ago, Bernard Bihari's daughter had a college friend who was newly diagnosed with MS. The girl suffered three severe exacerbations during the first year after her diagnosis and was hospitalized for two of these episodes. These were episodes of transverse myelitis in which the MS activity in her spinal cord mimicked that of a severed spinal cord. In two episodes the spinal cord was affected at waist level, causing movement and sensation disturbances in her legs and also impaired bladder and bowel function.

After each episode she experienced a quick recovery of 90–95 percent. The third attack affected a higher region of her spinal cord, in the region of the neck. Numbness spread throughout her body and terrified her. Again, she made a good recovery but lived in fear of another attack.

Bihari recommended that the girl try using 3 mg naltrexone at night. The girl stayed on LDN for several years and had no further exacerbations or deterioration in her condition. She stayed fairly active and, thinking her condition was in permanent remission, she stopped taking LDN. Several weeks later she experienced an exacerbation that caused severe muscle spasms in her left arm and extreme pain.

She again consulted Bihari and resumed taking LDN. Immediately, her arm improved and the spasm eventually stopped. Although she experienced some residual loss of nerve function in her arm, she was able to continue her active lifestyle.[10]

The Internet and MS

Once this account was posted online, it spread through cyberspace quickly. Around this time, Bihari and several of his colleagues in Manhattan began prescribing LDN for their patients with MS. As a result, more and more success stories appeared online, particularly on the MS bulletin boards and chat rooms. Patients were unwilling to accept no for an answer when they asked their doctors about LDN. Soon, there were stories (like the one described in chapter 9) of patients taking a proactive approach, abandoning canes and wheel chairs and resuming normal activities.

The fortunate patients who responded well to LDN weren't bragging when they told their stories. They were on a mission to help others learn about the potential benefits of LDN. Having to convince their doctors that LDN was worth trying, these early patient pioneers understood the need for clinical trials. Some of the trials described in the next section are a direct result of these patient efforts.

LDN Research Results

Worldwide, LDN is being studied more in MS than in any other disease. Anecdotal reports of large number of patients with MS effectively being treated with LDN by Bernard Bihari in the United States and Dr. Robert Lawrence in the UK have attracted researchers at universities as well as MS organizations. As a result, a number of clinical trials are underway, described in Appendix A.

Animal studies are important for establishing the safety and efficacy of LDN. In his research laboratory at Penn State, Ian Zagon and his team have been studying the effects of LDN in animals with an experimentally induced form of encephalomyelitis, which is essentially the same as multiple sclerosis in humans. In addition to demonstrating safety and efficacy, animal trials allow for tissue studies and are necessary for determining the detailed biological effects of LDN.

According to Yash Agrawal, human trials should include patients with RRMS as well as either SPMS or PPMS. The RRMS patients provide a sensitive indicator of the ability of LDN to reduce the number of relapses, while the SPMS/PPMS patients are better for monitoring disease progression.[11]

Beta Endorphins in MS

A study performed by Maira Gironi and her team in Italy shows that, in general, patients with MS have lower endorphin levels than those seen in patients who are disease-free. Patients with PPMS and SPMS were shown to have lower levels of beta endorphins than patients with relapsing forms of MS. Average beta endorphin concentrations were also lower, although not significantly, in patients with magnetic resonance-imaging enhanced lesions compared to patients without lesions. These data suggest that beta endorphins may have a role in regulating the inflammatory process.[12]

Evers Klinik, Sundern, Germany

At the First Annual Low Dose Naltrexone Conference in 2005, Dr. Zvonko Mir, presented the results of the first human trial of LDN for MS. Mir conducted this trial in October of 2004 at the Evers Clinic in Germany, a facility that specializes in the treatment of MS.

Before initiating this trial Mir reviewed more than 1,000 anecdotal reports from the United States, and forty reports from Germany. These reports suggested that LDN improved symptoms of spasticity, leg weakness, fatigue, and bladder urgency. He determined that the claims warranted investigation with a double-blind, randomized trial.

The clinical trial Mir designed involved sixty patients with PPMS and SPMS with expanded disability status scale (EDSS) scores of 5.0 + who were enrolled in the Evers Clinic as inpatients. Patients could not be on benzodiazepines such as diazepam or analgesics containing opiates. Half of the patients were treated with 3 mg LDN daily given at 9 A.M., for a duration of ten days. The remaining participants were given a placebo. Effects were measured by comparing pre-and post-neurological examinations and patient questionnaires.

The conclusions were:

• No significant differences between medication and placebo group
• LDN is not a generally applicable medication for the treatment of typical symptoms of MS
• Only ten patients in the drug group were counted as responders

However, in evaluating these results Mir discovered that those who found improvement with LDN responded very quickly. But overall, the study design was flawed. In the Evers trial he was investigating whether LDN caused an immediate reduction of symptoms in MS. He was specifically comparing LDN to treatments like Amantadine, which is used for fatigue, and Baclofen, which is used for spasticity. This trial also excluded patients with RRMS. The focus was on the progressive categories of MS with longer-term diagnosis (fifteen years average) who were likely to have higher degrees of disability (EDSS scores of 5+ or more).

The trial's conclusion also noted that "there must be some kind of effect," because, despite the double-blind design, 82 percent of the physicians and 75 percent of the patients were able to readily determine if the patient was administered LDN or the placebo. The researchers were also surprised to note that the responder group of ten patients improved their EDSS rating by 0.5 points in only ten days.[13]

Mir also noted that patients who responded to LDN, whom he continued to follow anecdotally for nine months, recommended that LDN be investigated in long-term trials. At the First Annual LDN Conference, other physicians in attendance, including Maira Gironi, agreed that a longer trial of LDN for PPMS was needed.

Pennsylvania State University Animal Trial

Experimental autoimmune encephalomyelitis (EAE), which was once referred to as experimental allergic encephalomyelitis, is a disease characterized by brain inflammation. EAE is an autoimmune, inflammatory, demyelinating condition. In humans, EAE primarily occurs as an acute condition of encephalomyelitis. EAE can also be induced in animals. In animals, EAE causes a condition that serves as a model for multiple sclerosis.

Animal models of EAE are induced in an effort to have the animals, usually rodents, develop either: 1) an acute condition that mimics acute disseminated encephalomyelitis in humans, or 2) a chronic condition that mimics multiple sclerosis in humans. Animal models allow researchers to study the biochemical effects of various therapies to gain an understanding of how these therapies work in specific diseases.

Ian Zagon at Penn State University was granted a Pilot Award by the National Multiple Sclerosis Society (NMSS) to study the effects of LDN in rats with induced conditions of chronic EAE. The study began on September 1, 2006, and ended on August 31, 2007. Titled "Role of Opioid Peptides and Receptors in MS," this study is considered the first step needed to validate or disprove the anecdotal reports of LDN's role in MS.

The Hypothesis

Zagon hypothesized that:

1. If continuous opioid receptor blockade (with high doses of naltrexome) exacerbates EAE it is because endogenous opioids are good for people with multiple sclerosis; the reason for exacerbation is that the continuous blockage would deprive subjects of endogenous opioids. However, if high doses of naltrexone and continuous receptor blockade cause improvement in EAE, it would show that endogenous opioids halt the disease process and that diminishing the interaction of endogenous opioids and their receptors would be beneficial.
2. If low doses of naltrexone and a short blockade of the opioid receptor helps to prevent EAE, it is because either the drug itself or the increase in endogenous opioids and their receptors caused by the drug is beneficial. If the effects are due to the drug itself, one would expect that a higher dose of naltrexone would be beneficial.
3. If low dose naltrexone exacerbates the disease, then endogenous opioids might be bad for EAE. In this case, blocking the opioid receptor continuously with high doses of naltrexone would cause improvement.

The Study

In the study, animal subjects with induced EAE were given naltrexone to determine its effects on the disease process. In addition, to determine if the effects were caused by naltrexone itself or were due to biological effects caused by naltrexone, animals were given either high or low doses of naltrexone. The importance here lies in observing effects related to either intermittent (low dose naltrexone) or sustained (high dose naltrexone) blockade of the opiate receptor. Thus, this research project also raises the question of whether endogenous opioids and opioid receptors influence the course of multiple sclerosis.

During the course of their study Zagon and his team set out to find the effects of naltrexone and endogenous opioid peptides on EAE. Zagon theorized that if endogenous opioids alleviated symptoms, then they are bad for the disease course of MS and, as a result, good for patients.

The results of this project are important in showing whether the complex formed by opioid growth factor (OGF) and the OGF receptor influences the disease course. Zagon has previously shown that a low dose of naltrexone produces a compensatory increase in native opioids and opioid receptors following the transient receptor blockade (see chapters one and five for more information on OGF and its receptor). While high doses of naltrexone prevent opioid receptor interaction throughout the entire day, low doses of naltrexone block the opioid receptor intermittently for four to eight hours; this is followed by a supersensitive reaction that occurs when naltrexone is no longer present (the remaining sixteen to twenty hours per day).

Study Results

The animals were evaluated with daily body weight measurement and EAE clinical scores. In addition, the animals were assessed weekly for immune response and outcome measures, and they were qualitatively and quantitatively evaluated on days ten (onset, before symptoms of EAE occurred), day twenty (behavioral symptoms expressed), and forty (chronic-behavioral symptoms expressed) for inflammatory index, demyelination, apoptosis, and for neuronal pathology.

The results of Zagon's completed animal study have demonstrated the underlying mechanism in which LDN affects disease progression in MS. In addition, the study provides evidence for the role of endogenous opioids and opioid receptors in MS. In summary, Zagon's animal study opens a new field of research related to disease development in MS as well as the role of opiate antagonists for the treatment of MS.

Although Zagon's animal trial results are now in the publication process and cannot be released prior to publication, the results suggest that continuous opioid receptor blockade exacerbates the progression of MS, whereas a low dose of naltrexone (and an increase in opioid growth factor) retards the disease course. The results show that a reduction in endogenous opioids is directly related to the onset and pathogenesis of an MS-like disease.[14]

Zagon is in the process of initiating clinical trials in MS and in Gulf War Syndrome that will be conducted together with a team of neurologists and other clinical specialists. Conditions similar to chronic fatigue syndrome such as those that comprise Gulf War Syndrome are suspected of being caused by increased nitric oxide synthesis.[15] Because of the role LDN plays in reducing nitric oxide (described further in chapter four), it is being investigated for a number of neurodegenerative disorders.

Clinical Trials of LDN in Human Subjects with MS

In 2007, neurology researcher Dr. Bruce Cree and his colleagues from the University of California, San Francisco, and Dr. Maira Gironi and her team from Instituto Scientifico San Raffaele in Milan, Italy, initiated two separate clinical trials of LDN in patients with multiple sclerosis. Both trials concluded that LDN offers benefits for patients with MS and that more studies are indicated. A substantial contribution toward the San Francisco LDN study was made by the grassroots efforts and contributions of MS patients through the LDN for MS Research Fund. Neither trial was funded by pharmaceutical companies.

Cree and Gironi presented descriptions and summaries of their trials in April 2008 at the 60th annual meeting of the American Academy of Neurology (AAN) in Chicago, Illinois. This event attracted more than 10,000 researchers and practicing neurologists from around the world.

Also, in early 2008 Dr. David Pincus from the MindBrain Consortium and the Department of Psychiatry of Summa Hospital System in Akron, Ohio, began conducting a study of LDN in patients with MS. The sixteen-week, double-blind, randomized, placebo-controlled, crossover-design study of thirty-six patients was held in conjunction with the Oak Clinic for the Treatment of Multiple Sclerosis. The study is examining symptom severity as well as any changes in quality of life, sleep patterns, and affective (emotional) states. Originally, trial results were expected in the summer of 2008.

However, although the subjects on LDN will be allowed to complete the trial, a flaw in the study design prevents meaningful results. Pincus plans to make changes to the protocol and begin a new trial of LDN in MS.

The San Francisco LDN Trial

Cree presented the results of his study "A Single Center, Randomized, Placebo-Controlled, Double-Crossover Study of the Effects of Low Dose Naltrexone on Multiple Sclerosis Quality of Life" on April 15, 2008. All patients in the study were given either 4.5 mg naltrexone or a placebo nightly for the first eight weeks. This was followed by a one-week washout in which no drugs were administered. Afterwards, subjects were switched to the alternate study drug for an additional eight weeks, giving all subjects an opportunity to receive treatment with LDN.

Trial Participants

Trial participants included eighty men and women with MS between eighteen and seventy-five years of age taking either glatiramer acetate, interferon beta, or no disease modifying treatment. Nine patients withdrew voluntarily, and one withdrew secondary to an unrelated medical condition. Seventy patients completed both treatment periods. Of these, sixty patients provided complete data, and the other ten subjects could not be included because of incomplete data, including four data management errors and six uncompleted surveys.

Most patients had relapsing remitting multiple sclerosis (RRMS) followed by primary progressive multiple sclerosis (PPMS), and secondary progressive multiple sclerosis (SPMS). One patient had progressive relapsing multiple sclerosis (PRMS).

Inclusion criteria:

• Clinically definite MS diagnosed by current International Panel Criteria
• Whether on current therapy or not (approved therapies include either interferon beta or glatiramer acetate, but not both, with continuous use for three months or longer), a willingness not to change or start new disease modifying or symptomatic therapies of MS during the course of the seventeen-week trial
• For women of childbearing potential, willingness to use a barrier method during the course of the study

Exclusion criteria:

• Starting a disease modifying therapy within three months of entry in the trial
• Planned start of disease modifying therapy during the clinical trial
• Pregnancy
• Current chronic opioid agonists use
• Patients currently on both interferon beta and glatiramer acetate

- Patients currently taking LDN (before the start of the study)
- Patients currently taking immunosuppressive medications including cyclophosphamide, mitoxantrone, azathioprine, methotrexate, mycophenolate mofetil, natalizumab, rituximab, alemtuzumab or other immune suppressants
- Participation in other clinical treatment trials in MS
- Patients who cannot comprehend MSQLI54 instructions and are unable to complete the evaluation forms
- Patients under eighteen years of age
- Patients older than seventy-five years prior to the start of therapy

UCSF Trial Hypothesis

Lesions in MS may be the result of oligodendroglial apoptosis and microglial activation rather than neuroinflammatory processes. Activated microglial cells secrete proinflammatory and neurotoxic factors (nitric oxide and peroxynitrites) that could cause neurodegeneration. In theory, inhibition of microglia would be protective in MS. At low doses, naltrexone is capable of reducing microglial cytokine (IL-1β) and nitric oxide in glial cultures. If naltrexone is beneficial in MS, a possible mechanism of action is through reduction in microglial nitric oxide synthase activity resulting in decreased peroxynitrite production. Peroxynitrites are thought to inhibit glutamate transporters thereby increasing the synaptic concentrations of glutamate resulting in excitatory neurotoxicity.[16]

UCSF Trial Results

All patients in the San Francisco LDN study were evaluated using the Multiple Sclerosis Quality of Life Inventory (MSQLI54) and the visual analog scale (VAS) prior to starting the trial (baseline level) and at weeks eight and seventeen. The only baseline covariate that had a statistically significant impact on the model was the baseline score. A statistical analysis of the results showed that the benefits of LDN were not affected by the disease course, age, treatment order, or treatment with either interferon beta or glatiramer acetate. The primary outcome measure of the trial was based on a comparison of the mean score between treated and placebo groups, adjusting for differences in the baseline score between the two groups—comparison of each of the ten scales of the MSQLI54 between treated and placebo groups.

LDN was shown to have greatest efficacy in mental health parameters, including improved wellbeing, and in parameters of mental health status, based on the Mental Health Status Summary Scale. The improvement in pain based on the pain effects scale was also significant. Pain assessment by the perceived deficits questionnaire was also significant. Other parameters evaluated in the study included social support, fatigue impact, sexual satisfaction, and physical components.

The study did not attempt to assess the potential effects of LDN beyond eight weeks of treatment. The short duration of the study may have been inadequate to evaluate all of the components studied. It should be noted that many of the patients did not have sexual partners and could not adequately be evaluated for sexual satisfaction.

Study Conclusions

1. Eight weeks of treatment with LDN significantly improved quality of life indices for mental health, pain, and self-reported cognitive function of MS patients as measured by the MSQLI.
2. An impact on physical quality of life indices including fatigue, bowel and bladder control, sexual satisfaction, and visual function was not adequately tested for and, thus, was not observed.
3. The benefits of LDN were not affected by disease course, age, treatment order, or treatment with either interferon beta or glatiramer acetate.
4. The only treatment related adverse event reported was vivid dreaming during the first week of the study; this was noted in some, but not all, patients given LDN.
5. Potential effects of LDN beyond eight weeks of treatment were not addressed in this study.
6. Multicenter clinical trials of LDN in MS are warranted.[17]

Patient Reports

Several MS patients who participated in the study reported noticing significant improvement in nighttime bowel and bladder function after starting LDN. Several patients also reported withdrawing from the study after the first eight weeks because they felt so good they didn't want to stop taking LDN.

University of Milan, Italy

In late 2006, Dr. Maira Gironi began a six-month trial of LDN that focused on the hardest-to-treat MS patients, those with primary progressive multiple sclerosis (PPMS). To date this is the only LDN trial for MS that includes biochemical evaluations of beta-endorphin levels in addition to tracking effects on symptoms (spasticity, pain and fatigue). It is also the only trial so far to study the effects of LDN in isolation from other MS immunosuppressive or immunomodulator drugs. Patients were allowed to remain on gabapentin and selective serotonin reuptake inhibitors but not opiates, immunosuppressive or immunomodulator drugs. The study was supported by the Italian Federation of Multiple Sclerosis.

In October 2007 Gironi presented an initial report at the European Congress of MS in Prague. Data showed that the study involved forty patients with a diagnosis of PPMS (nineteen males, twenty-one females, mean age 53.4 years, mean age at disease onset 41.2, median expanded disability status scale [EDSS] score of 6.0 with a range between 3.0 and 6.5).

The study was described as a six-month, multi-centric, open-label pilot study intended to evaluate safety and efficacy on spasticity, pain and fatigue. The study included four weeks of prescreening, twenty-four weeks of treatment and four weeks of follow-up. Patients were initially given 2 mg naltrexone daily (in the morning) for two weeks and then the dose was increased to 4 mg.

Milan Trial Results

The study concluded that LDN was safe and effective at recommended doses. Evaluation was performed with appropriate testing including laboratory tests, Ashworth Scale, Fatigue Severity Scale, and the Visual Analogue Scale.

SIDE EFFECTS. Transitory hematological abnormalities (increased liver enzymes, hypercholesterolemia), mild agitation, urinary infection, and sleep disturbance were the most common adverse events reported. Only five patients withdrew from the study—one for protocol violation, one for severely increased hypertonia (muscle tension), one for nocturnal urinary incontinence related to deeper sleep, one for an increase in total and indirect bilirubin, and one for urinary tract infection.

QUALITY OF LIFE ISSUES. In regard to quality of life, there was a general trend of improvement observed between the final and baseline evaluations. The peak of efficacy was seen at the 3-month visit. The most beneficial effects seen were improvement in fatigue, depression, physical functioning, energy, and social functioning.

BETA ENDORPHIN LEVELS. Test subjects showed increased levels of peripheral blood mononuclear cell (PBMC) beta-endorphins as early as three months after beginning therapy. At six months of LDN therapy, PBMC beta endorphin elevations became statistically significant compared to baseline levels.

This increase was still statistically significant one month after therapy was discontinued.

This study summary states that patients with MS show low PBMC beta endorphin levels. It further states that the traditional functions of beta endorphins include modulation of pain, mood and endocrine secretion, and that the immunomodulating functions of beta endorphins include: inhibition of antigen induced T-cell proliferation, down regulation of proinflammatory cytokines, and inhibition of macrophage IL-6 and IL-12 production.[18]

4

LDN in Neurodegenerative Disorders

For more than a decade doctors have been prescribing low dose naltrexone as an off-label treatment for patients with neurodegenerative disorders such as Parkinson's disease and multiple sclerosis. Many of these patients have experienced a halt in disease progression, and some patients have noticed considerable improvement. This chapter focuses on the role of LDN in reducing disease progression in neurodegenerative disorders.

Animal studies of neurodegenerative diseases show that LDN has antioxidant properties and reduces inflammation related to persistent immune system activation. Pilot human trials of LDN have been completed and early trial results indicate a need for phase 2 trials. Anecdotal evidence suggests that LDN holds promise for several neurodegenerative diseases. Tony White, a retired captain in the U.S. Navy who has Parkinson's disease, has agreed to share the story of his success with LDN. It can be found at the end of this chapter.

What Are Neurodegenerative Diseases?

Neurodegenerative diseases are members of a family of progressive neurological disorders. Signs and symptoms of neurodegenerative disorders are caused by the deterioration of central nervous system cells (neurons) or the protective myelin sheath covering nerve fibers. Damage to these cells and tissues disrupts transmission of the cellular signals that control motor function, cognitive function, and affect. Consequently, neurodegenerative disorders cause disturbances in one or more of these functions.

Similarities and Differences

Some neurodegerative disorders have common features. For instance, ataxia (disturbance in movement) is seen in multiple sclerosis and amyotrophic lateral sclerosis (ALS), whereas cognitive disturbances are seen in Alzheimer's disease and Parkinson's disease. However, the primary pathological abnormalities which contribute to disease and account for the specific neurological symptoms in these disorders may have different origins.

Causes of Neurodegeneration

Increasing evidence suggests that inflammation in the brain contributes to the underlying disease process in several neurodegenerative disorders, including Alzheimer's disease, amyotrophic lateral sclerosis, Parkinson's disease, multiple sclerosis, and AIDS dementia.[1] In addition, changes in structural proteins caused by genetic mutations contribute to the disease process in Alzheimer's disease and Parkinson's disease.

In the neurodegenerative disorder multiple sclerosis (MS), increasing evidence confirms previous reports suggesting that microglial activation is the primary cause of myelin deterioration and the development of brain lesions.[2]

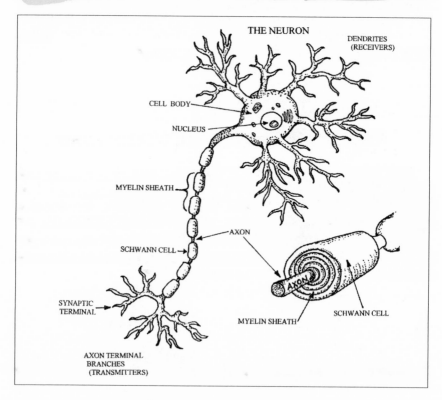

The neuron (Marvin G. Miller)

Brain Inflammation

Tissue studies of plaque from the brains of patients with Alzheimer's disease suggest that brain inflammation is the underlying cause. These studies also show that inflammatory cytokines, complement proteins, and microglial cells are contributing to neurodegeneration rather than occurring in response to the degenerative process.[3]

Immune Function and Microglial Activation

Glial cells, which include microglia and astrocytes, are the resident immune system cells of the central nervous system. According to current theory, microglia help clear dying cells that are ready for elimination. Microglia also assist with the immune surveillance and host defenses characteristic of immune system cells. Microglia show particular sensitivity to changes in their microenvironment. Thus, they become activated readily in response to infection or injury.

Activation of glial cells, especially activation of microglial cells, is the hallmark of brain inflammation. Activated microglia have been found in the brains of patients with Alzheimer's disease, Parkinson's disease, HIV/AIDS dementia complex, amyotrophic lateral sclerosis, multiple sclerosis, and prion-related diseases. In addition, tissue studies performed postmortem on the brains of patients with neurodegenerative diseases show evidence of the involvement of microglia in neurodegeneration.

Production of Neurotoxic Factors

Glial cell activation triggers the production of a variety of pro-inflammatory and neurotoxic (destructive to neurons) factors, including several cytokines such as tumor necrosis factor-alpha (TNF-α) and interleukin-1β (IL-1β); fatty acid metabolites; and free radicals, such as nitric oxide and superoxide. Studies show that the combination of TNF-α and IL-1β, but not either cytokine alone, induces neurodegeneration of neurons in the cortex of the brain.[4] Often referred to as the gray matter, the cerebral cortex is essential for perception, memory, awareness, attention, thought, language, and consciousness.

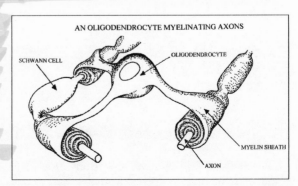

Oligodendrocyte myelinating axons (Marvin G. Miller)

PROTECTIVE FACTORS. Although microglia primarily produce neurotoxic factors, they also secrete a small quantity of protective factors, including the glia-derived neurotrophic factor, which is potentially beneficial to the survival of neurons. Production of neurotrophic factor causes an effect similar to the neuroprotective function of activated astrocytes.

Nitric Oxide

As mentioned above, microglial activation causes the production of several free radicals including nitric oxide (NO), superoxide, and peroxynitrite molecules. Nitric oxide is produced by the catalytic action of the enzyme nitric oxide synthase (NOS). NOS causes the amino acid arginine to react with oxygen to form citrulline and NO.

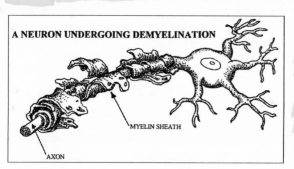

Neuron undergoing demyelination (Marvin G. Miller)

Three distinct types of nitric oxide synthase have been discovered: endothelial cell-derived NOS, neuronal NOS, and the inducible NOS (iNOS). During microglial activation, NOS is induced, and it is this iNOS that contributes to excess NO and peroxynitrite molecules. Research shows that NO has several diverse physiological functions such as muscle relaxation, neuronal activity, and immune modulation.[5]

EXCESS NO IN BRAIN SIGNALING. Nitric oxide is also necessary for certain biochemical processes involved in brain signaling. However, in excess, as seen in microglial activation, nitric oxide is detrimental to health. The excess peroxynitrites that result from microglial activation are reported to induce neurodegeneration by inhibiting glutamate transporters, which, in turn, leads to excess production of the excitotoxin glutamate. In excess, glutamate destroys neurons.

Consequences of Microglial Activation

In summary, microglial activations induces production of the enzyme nitric oxide synthase (iNOS). The iNOS causes excess production of NO, peroxynitrites, and superoxide molecules.

As a free radical, NO is one of the major contributors of reactive nitrogen species. This allows NO to behave in two different ways: first as an antioxidant scavenging oxygen free radicals and protecting cells from reactive oxidative insults (similar to rusting on a car), and, second, directly reacting with proteins, especially iron-containing enzymes, such as guanylate cyclase, cytochrome P450 (necessary for drug metabolism), and cyclooxygenase (which leads to inflammation), to modulate their functions.

In addition, NO can react with the lipid (fatty) surface membrane of cells to induce lipid peroxidation, a process that causes the accumulation of melanin deposits on hands and other signs of aging. Indirectly, the combination of NO and superoxide molecules can form highly reactive intermediate compounds, such as peroxynitrite, that can induce DNA strand breaks (mutations), lipid peroxidation, and protein nitration.

Reactive Gliosis

Inflammation in the brain primarily involves the participation of the two major types of glial cells: the microglia and the astrocytes. Normally, microglia guard and protect the immune system, and astrocytes act to maintain ionic homeostasis (cellular fluid balance), buffer the action of neurotransmitters such as glutamate, and secrete nerve growth factor to ensure neuronal regeneration. However, when glial cells are activated by injury or immunologic challenge, a process of activation known as reactive gliosis, including activation of both microglial cells and astrocytes, occurs.

Reactive gliosis has been observed as part of the disease process in Alzheimer's disease, Parkinson's disease, multiple sclerosis, AIDS dementia complex, cerebral stroke and traumatic brain injury.

Activated astrocytes secrete neuotrophic growth factors in an attempt to enhance the survival of neurons. Astrocytes secrete small amounts of NO and IL-1β to help destroy invading pathogens. Activated microglia lead to excess production of reactive nitrogen species. In reactive gliosis, however, the accumulations of NO and cytokines are damaging to neurons and lead to persistent inflammation and neurodegeneration.

Other Neurotoxins

Several other neurotoxins are known to directly contribute to neurodegeneration both on their own and because of their ability to induce microglial activation. These include beta-amyloid peptides in Alzheimer's disease, HIV coat protein gp120 (a cell surface protein that contributes to HIV/AIDS dementia), prion protein-derived peptides, and the pesticide rotenone. Neuronal damage may also occur directly from ischemia (lack of oxygen) and mechanical injuries.

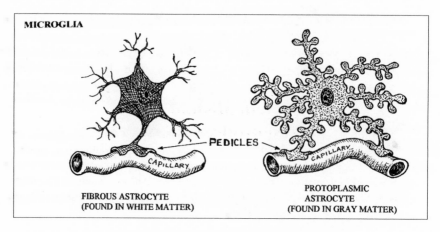

Macroglia (Marvin G. Miller)

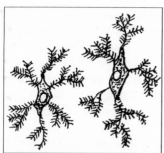

Microglial Activation and Opiate Antagonists

Opiate antagonists, such as naloxone and naltrexone, are currently being investigated for their ability to inhibit microglial activation and prevent neurodegeneration. In an animal model of Parkinson's disease, naloxone was able to reduced lipopolysaccharide-induced microglial activation.[6] This observation has prompted additional studies on microglial activation, which are described later in this chapter in the section on Parkinson's disease.

The opiate antagonist naltrexone used at low doses may inhibit nitric oxide synthase activity. This, in turn, may decrease the formulation of peroxynitrites, which has the potential to decrease glutamate neurotoxicity. Dr. Priti Patel, director of the St. John's University College of Pharmacy's Drug Information Center in Queens, New York, describes this

OLIGODENDROGLIA
(ATTACHED TO NEURONS)

Top: Microglia. *Bottom:* Oligodendroglia (Marvin G. Miller)

hypothesis and explains how low dose naltrexone could benefit patients with multiple sclerosis in a letter to the editor of the *Annals of Pharmacotherapy*, in which he emphasizes the need for clinical trials of naltrexone.[7]

Protein Misfolding
and Neurodegeneration

Errors in protein folding also lead to neurodegeneration. In oxidative stress, free radicals can trigger injurious pathways or post-translational modifications that cause alterations in the formation of various proteins, including enzymes. This misfolding of protein can lead to abnormal protein accumulations and mimic genetic mutations. In Parkinson's disease, protein misfolding of the enzyme parkin leads to abnormal protein accumulation, directly contributing to pathology. Changes related to protein misfolding induce neuronal cell injury and death.[8]

Causes of Microglial Activation and Protein Misfolding

Suspected causes of microglial activation and protein misfolding include excess overstimulation (excitotoxicity) of glutamate receptors and oxidative stress. Excitotoxicity, which is described below, has been implicated as a final common pathway contributing to neuronal injury and death in many acute and chronic neurological disorders, including Parkinson's disease, amyotrophic lateral sclerosis, multiple sclerosis, Alzheimer's disease, stroke, and trauma.

Neurotransmitters

Neurotransmitters are chemicals that send messages or signals allowing cells to communicate with one another. Neurotransmitters normally attach to receptors on the surface membrane of nerve fibers, which causes a change in the membrane's shape that allows for the passage of sodium and calcium molecules. This triggers the cell to fire and transmit a signal down its axon fibers.

Glutamate

Glutamate is the most abundant neurotransmitter found in the brain. Glutamate's role is primarily that of an excitatory substance. Many areas of the brain contain glutamatergic neurons, including the cerebral cortex, striatum, hippocampus, hypothalamus, thalamus, cerebellum, and the visual and auditory system.

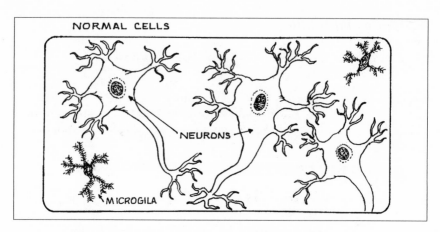

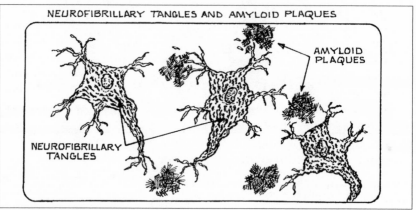

Top: Normal neurons (Marvin G. Miller). *Bottom:* Neurofibrillary tangles and amyloid plaques (Marvin G. Miller).

Glutamate Receptor Subtypes

There are at least twenty different cell receptors that glutamate and related molecules can bind to. The glutamate-like substance N-methyl-D-aspartate (NMDA), for instance, only binds to glutamate subtypes known as NMDA receptors. A second substance, quisqualate, only stimulates receptors of the glutamate subtype quisqualate. Another glutamate receptor, the kainite receptor, only reacts with the chemical kainite.

Glutamate itself, in the form of monosodium glutamate (MSG) and aspartate can stimulate all of the three major types of glutamate receptors. However, glutamate shows some selectivity in choosing neurons from certain areas of the brain and skipping neurons in other areas.

Excitatory and Inhibitory Balance

When glutamate and aspartate molecules activate glutamate receptors they influence brain systems concerned with sensory perception, memory, orientation in time and space, cognition, and motor skills. Normally, the brain depends on a careful balance between excitatory systems such as glutamate and inhibitory systems (positive and negative impulses). Disruption of this balance can lead to neurological symptoms ranging from a fine tremor of the hands to a full-blown seizure.

The NDMA Receptor

The NDMA receptor is the most common of the three major types of glutamate receptors found on neurons in the brain. The NDMA receptor serves as a lock, which, when activated, opens a calcium channel that regulates the entry of calcium into the inside of the neuron. Neurons also have receptors that allow for the passage of zinc, magnesium, and the amino acid glycine. These receptors must also be activated in order for glutamate to fire nerve cells. Glycine causes the neurons to become more sensitive to activation by excitotoxins such as glutamate. If the nerve stimulation is persistent and unrelieved the neurons are eventually destroyed.[9]

Calcium and Protein Kinases

Excitotoxins such as glutamate and aspartate work by binding with glutamate receptors and opening calcium channels on certain receptor subtypes. When these receptors are persistently activated, the calcium channel becomes stuck in the open position, which allows a large influx of calcium to pour into the cell.

Within the cell, calcium activates an enzyme known as protein kinase C. This enzyme causes the release of calcium from the cellular storage site known as the endoplasmic reticulum. Excess levels of calcium cause the release of the fatty acid arachidonic acid.

The presence of arachidonic acid, in turn, stimulates the release of lipoxygenase and cylco-oxygenase enzymes, which are released in a homeostatic attempt to destroy the acid. During the process, free radicals are produced. The final chain of events resulting from excitotoxicity leads to a cycle of destructive reactions that contributes to cell death.

Excitotoxins and Opiate Antagonists

Researchers at the National Institutes of Health have shown that morphinan compounds such as dextromethorophan and naltrexone prevent glutamate toxicity in neurodegenerative disorders by inhibiting the NDMA receptors on neurons.[10]

LDN's Role in Preventing Excitotoxicity

It's been proposed that naltrexone acts by reducing the production of inducible nitric oxide synthase.[11] This decreases the formation of peroxynitrite molecules, which, in turn, prevents the inhibition of the glutamate transporters. No longer inhibited, the transporters move glutamate from the circulation preventing excess glutamate from accumulating. This prevents the excitatory neurotoxicity on neuronal cells and oligodendrocytes caused by excess glutamate.

Reducing excitotoxicity reduces microglial activation and it reduces apoptosis (programmed cell death) of the myelin-producing oligodendrocytes. This, in turn, prevents destruction of the myelin sheath covering nerve fibers. Destruction of the myelin sheath is the underlying pathology in MS.

Parkinson's Disease

Parkinson's disease is one of the most disabling of the chronic neurodegenerative diseases. Studies show that Parkinson's disease is associated with a significant loss of quality of life. Medications such as dopamine agonists, amantidine, and other drugs are effective at reducing symptoms and increasing mobility, but they typically fail after several years. Deep brain stimulation of the subthalamic nucleus has shown to be more effective than chemical treatment alone for improving motor neuron symptoms. However, because it is an invasive procedure, this treatment is more likely to cause serious adverse effects, including hemorrhagic stroke.[12]

Symptoms

Symptoms in Parkinson's disease vary in severity and predominant symptoms can change over time. Symptoms that can occur in Parkinson's disease include tremor, rigidity, a slowness of movement known as bradykinesia, an absence of movement known as akinesia, balance problems and postural instability, abnormal muscle contractions (dystonia), lowered blood pressure upon rising (orthostatic hypotension), gait disturbances, speech disturbances, breathing disturbances, muscle pain, joint pain, sleep disturbances, digestive disturbances, weight loss, dizziness, difficulty swallowing (dysphagia), mood disturbances, apathy, cognitive changes, short-term memory loss especially for procedural tasks, urinary incontinence, and dementia.

Disease Process

Parkinson's disease is characterized by degeneration of the pigmented dopamine-producing (dopaminergic) neurons in the pars compacta region of the substantia nigra area of the brain. The dopaminergic neurons project to the striatum. In Parkinson's disease there is insufficient dopamine to stimulate cells, causing alterations in the activity of the neural circuits that regulate movement. The overall effect is an inhibition of the direct dopaminergic pathway and excitation of the indirect dopaminergic pathway.

The brain has four major dopamine pathways. The pathway that mediates movement is the first pathway affected in Parkinson's disease. Other pathways govern volition, emotional responsiveness, desire, initiative, reward, sensory processes, and maternal instincts. Disruption of these pathways accounts for the symptoms that occur in Parkinson's disease. The Unified Parkinson's Disease Rating Scale is the primary clinical tool used to assist in diagnosis, determine severity of PD, and monitor the response to therapy.

Neuronal Damage

Various theories have been proposed to explain why dopaminergic neurons undergo degeneration in Parkinson's disease. The primary cause is an abnormal accumulation of the protein alpha-synuclein, which is bound to ubiquitin in damaged dopaminergic neurons. This accumulation of protein forms inclusions called Lewy bodies that lodge in the cell cytoplasm. Accumulations of iron and copper are also found in damaged dopaminergic neurons. This suggests that free radicals contribute to the underlying neuronal damage in Parkinson's disease.

Several genetic mutations have been found that are suspected of contributing to a familial form of Parkinson's disease. An increased susceptibility to the toxic effects of certain chemicals, particularly pesticides and excitotoxins, is also suspected of contributing to neuronal damage in Parkinson's disease.

Parkinson-like Diseases

Patients with no evidence of other primary diseases that may be contributing to neurodegeneration are said to have primary Parkinson's disease. Several other conditions that closely resemble Parkinson's disease are called Parkinson-plus diseases. These include multiple system atrophy, progressive supranuclear palsy and corticobasal degeneration. In addition, the disorder Lewy body dementia, a disease similar to Alzheimer's disease is characterized by the presence of Lewy bodies in neurons, suggesting an association with Parkinson's disease. In general, the Parkinson-plus diseases progress more rapidly than primary Parkinson's disease.

Microglial Activation
in Parkinson's Disease

The involvement of microglial activation in Parkinson's disease was initially postulated as a result of postmortem studies of the brains of patients with Parkinson's disease. In these studies, large numbers of microglia with human leukocyte antigen-DR (HLA-DR) immune system markers were found in the substantia nigra region, the region of the brain primarily affected in Parkinson's disease. Other studies have confirmed these findings and shown elevated levels of proinflammatory cytokines and evidence of oxidative stress mediated damage. Studies have also shown that brain injuries early in life correlate to an increased incidence of Parkinson's disease in later life.

In addition, the substantia nigra region of the brain has been found to have a reduced antioxidant capacity, evidenced by low levels of the antioxidant glutathione. Normally, in brain injuries, natural supplies of cellular antioxidants are depleted in their efforts to prevent damage. Because of its reduced antioxidant capacity, the substantia nigra is particularly vulnerable to insult, including oxidative stress. Animals studies performed by Dr. Jau-Shyong (John) Hong and his team at the National Institute of Environmental Health Sciences (NIEHS) Division of the National Institutes of Health (NIH) have confirmed these findings.[13]

Opiate Antagonists
in Parkinson's Disease

In another animal study, Hong and his colleagues demonstrated that low doses of the opiate antagonist naloxone reduced microglial activation and offered neuroprotection. This was also seen in cell cultures and attributed to the inhibition of superoxide. This confirmed the notion of Hong that reducing the production of neurotoxins caused by microglial activation with naloxone offered protection against neurodegeneration. The mechanism by which naloxone works appears to be independent of the opioid receptor system and directly related to inhibition of superoxide production.[14]

High Dose Naltrexone in Parkinson's Disease

The effects of naltrexone and related opiate antagonists differ markedly depending on the dose, which in turn determines the duration in which the opiate receptor is blocked. High doses of naltrexone have been shown to stimulate cell proliferation, whereas low doses inhibit cell proliferation and reduce inducible nitric oxide synthase activity and its resultant excitatory neurotoxicity.[15] Studies using low doses of opiate antagonists in Parkinson's disease are currently underway at the NIEHS and other institutions.

A French study using high doses of naltrexone (100 mg/day) predictably failed to modify motor symptoms in patients with Parkinson's disease. In this one-month long double-blind randomized study, ten patients treated with high doses of naltrexone exhibited no change in motor function.[16] The high doses of naltrexone used in this study cause the opiate receptor to be persistently blocked. Therefore, the benefits related to low dose naltrexone and intermittent blockade would not have occurred.

Ian Zagon has found that LDN blocks the opiate receptor for three to four hours, after which endorphin levels increase dramatically for twenty-four to seventy-two hours. When the receptor is blocked continuously with high dose naltrexone, there is no increase in endogenous endorphins, and none of the benefits associated with LDN occur.

Alzheimer's Disease

Alzheimer's disease, which is the most common cause of dementia, is considered the prototypical neurodegenerative disease. In Alzheimer's disease, plaques containing deposits of beta-amyloid protein build up between neurons. Tangles containing twisted fibers of tau protein form inside dying neurons. These abnormal changes in neurons are thought to contribute to neurodegeneration.

In their continuing studies of microglial activation and neurodegeneration, John Hong and his team at the NIEHS have found that beta amyloid peptide (Abeta 1–42) exhibits enhanced neurotoxicity toward both cortical and mesencephalic neurons through the activation of microglia and the production of superoxide.

Naloxone in Alzheimer's Disease

The purpose of the NIEHS study of beta amyloid protein was to determine whether naloxone isomers had any effect on Abeta 1–42-induced neurodegeneration. Studies of midbrain cell cultures showed that naloxone and its inactive stereoisomer protected the dopaminergic neurons with equal potency. Naloxone inhibited lipopolysaccharide-induced microglial activation and the release of pro-inflammatory factors and superoxide free radicals.[17]

Anecdotal Accounts

As of September 2003, the neurologist Bernard Bihari, who had recently retired, reported on his Web site that he had seven patients in his practice with

Parkinson's disease who had shown no disease progression since starting LDN. Two of these patients were reported to have shown clear improvement in disease signs and symptoms. In addition, Bihari reported being contacted by three patients with Parkinson's disease who were treated with LDN by their personal physicians. Two patients showed improvement in their breathing, as measured by forced vital capacity (FVC), within several months, and the third patient experienced significant subjective improvement in his breathing ability and a significant reduction in his resting pulse.

Bihari also reported consulting with two patients with amyotrophic lateral sclerosis (ALS) who experienced benefits using LDN. Information from Australia on patients with ALS and primary lateral sclerosis (PLS) who are successfully being treated with LDN is also available on Bihari's Web site.[18]

Destiny Marquez, a patient advocate, is writing a book to describe the improvement her father has experienced using LDN for Parkinson's disease, which was exacerbated by a major hemorrhagic stroke related to deep-brain-stimulation therapy. Marquez reports that her father has been able to reduce his other medications for Parkinson's disease considerably by using a combination of LDN and hyperbaric oxygen therapy. He is now able to accompany her when she speaks to Parkinson's disease groups and show his progress first-hand.

Antibiotics and Naltrexone

Studies show that β-lactam antibiotics, such as penicillin and ceftriaxone, offer benefits in treating neurodegenerative diseases such as amyotrophic lateral sclerosis (ALS) and Huntington's disease. These antibiotics do so by selectively inducing transcription of the gene encoding one of the glutamate transporters. The neurotransmitter glutamate is essential for normal signal transmission between neurons, including the motor neurons, which trigger muscle contraction. It is the premature death of motor neurons that produces the progressive paralysis characteristic of ALS. Excessive glutamate, described earlier in this chapter, causes excitotoxic injury to neurons, which occurs in many neurological disorders, including ALS.[19]

The mechanism in which β-lactam antibiotics work in regulating the glutamate transporter parallels the action of low dose naltrexone. In both therapies, the glutamate transporter protein recovers synaptic glutamate in a manner that rapidly silences glutamate neurotransmission. Consequently, the neurons are protected from excess stimulation.

Yash Agrawal reports that studies of β-lactam antibiotics have led to a call for clinical trials of ceftriaxone in ALS. Ceftriaxone is also used to treat Lyme disease, a disorder that may coexist or masquerade as MS, another disorder characterized by glutamate excitotoxicity. In a letter to the editor of the *European Journal of Neurology*, Agrawal postulates that the improvement some patients with MS experience may be due to a reduction in excitotoxins rather than from the presumed eradication of borreliosis. Agrawal asserts that LDN may have the same effect as ceftriaxone. Because the long-term use of antibiotics can cause antibiotic resistance, the addition of LDN to an intermittent LDN protocol might offer several advantages. Agrawal concludes, "It is hoped that future clinical trials will explore these possibilities."[20]

Tony White and Parkinson's Disease

Tony White has given us permission to share his story of using LDN for Parkinson's disease. The following excerpts are taken directly from the account he provided.[21]

TONY WHITE'S STORY. My name is Tony White. I am sixty-seven years old and was diagnosed with Parkinson's disease in 1997. My first symptoms were tremors in my left hand and foot, and a limp, favoring my left side.

After graduating from the University of Washington, having a successful career in the U.S. Navy, and founding two large corporations, in early 2004 I resigned and retired. Parkinson's had taken control of my life and I simply could not continue. If I needed to walk more than a hundred yards, I would just freeze up and would need a wheelchair. I was also at a point where my tremor was so bad that I would either spill most of my food or, if it was a sandwich, literally throw it across the table. That meant that I could not eat out in public anymore.

Then in early 2006 I met a friend who told me about deep brain stimulation (DBS) surgery. My friend wrote, directed, and starred in a popular hourlong video of his own DBS brain surgery, which involves inserting two six-inch, stainless-steel, electrically charged rods into the subthalamic nucleus region of the brain.

Convinced of its benefits, I had DBS surgery performed. Initially, I could walk for miles and my terrible tremor and freezing symptoms were completely gone! It was a true modern medical miracle. Although there are a number of theories about how DBS works, the fact is, no one really knows why. It just does.

Unfortunately, for some of us there have been some negative side effects. For me, the side effects included serious balance problems and slurred speech. That, combined with the insomnia that most people with Parkinson's disease taking handfuls of Sinemet must endure, caused my life to remain very difficult.

Although I could walk long distances and eat in public again, I was falling and badly bruising my ribs and other vulnerable body parts almost daily. And I rarely got a good night's sleep. Certain that my probe placement was off, I consulted with my doctor and discovered that I had superb placement. However, I was not able to find a competent local programmer for almost two years.

My wife Charlene and I attended the World Parkinson's Congress in Washington DC in February 2006 and spent the day with DBS technicians trying to fix my transmitter problem. For the next eighteen months we continued to pursue solutions for this problem. In the meantime, my primary care physician tried his best to find a prescription drug that might get me the rest that I needed so badly. Unfortunately, those drugs either made me so groggy that I was completely dysfunctional the next day or, they didn't work for me at all.

During the summer of 2006, Charlene and I flew several times to the Cleveland Clinic in Ohio. The neurologist we saw was stumped by the severity of my imbalance problem and my post DBS unresponsiveness to Sinemet. She had me chew two 200/50s of Sinemet and nothing happened. Thinking the probes might be off, she ordered a special MRI for me. In fact, the very best DBS surgeons in the world examined that MRI, and they all confirmed that both of the placements were nearly perfect.

So again we consulted the DBS probe manufacturer. He recommended that we see a neurologist in Spokane, Washington. Within a few days we were in Spokane meeting with specialists, and within an hour I was tuned by a world-class DBS programmer and running wind sprints in the hallway. Everybody in that part of the building came out to the hall area to see what all the commotion was about.

Unfortunately, none of the prescribed remedies could rid me of my chronic insomnia. We tried 5 mg of Clonazepam and it seemed to be effective for a few days, but nothing could do a better job than plain old green Nyquil.

Then, in December 2006 Charlene and I went to a meeting with a local medical doctor–inventor who was looking for investors. Samantha (SammyJo) Wilkinson was there. The meeting itself was a disaster, but SammyJo mentioned low dose naltrexone (LDN) and her Web site. I knew that I had to follow up with her. It was as if I had known her my whole life.

I also ordered Mary Bradley's book *Up the Creek with a Paddle* and read it cover to cover within twenty-four hours. I then went to SammyJo's Web site and read every scrap of information that I could find about LDN.

The next day I called my doctor and discussed the use of LDN with him. I told him that I wanted to come to Spokane as soon as possible to see how quickly I could start the program. He had certainly done his homework and must have ordered my prescription before I arrived. He said that he was already familiar with LDN, that there had been a lot of research done on the use of Naltrexone at 50 mg/day and that although it is a very powerful drug at that amount, at the 4.5 mg/day level it was almost as safe as homeopathic medicine would be. He said outright that he wasn't certain whether or not LDN would help me much but he was very certain that it would do me no harm.

My doctor had me titrate the drug, starting low and increasing in small increments. I started out using 1.5 mg and its effect was immediate. On the very first evening that I used it, within ten minutes after taking the LDN, I was sound asleep. And for the first time in years, I slept all night. And the next. And the next. And the next!

For me it was sixty days of heaven, until I got up to a dose of 4.5 mg/day. Then, I guess that the dose had gotten high enough that the stimulant characteristics of the Naltrexone had kicked in. SammyJo recommended I see the leader of our forty-member PD Support Group, who was by this time also taking LDN and who told us that she took 100 mg of 5HTP (an over-the-counter Melatonin precursor) to combat the stimulant nature of the Naltrexone. She suggested that we try 50 mg to see if that might be enough to do the trick. I tried it and it didn't work, but the next night I took 100 mg and it did work.

Now, I'm sleeping at least eight hours every night. I feel rested in the morning for the first time in years. I take no naps during the day and have no inclination to do so. I complete a strenuous hour and a half workout regimen nearly every day of the week. And my balance is coming back. I can now balance on one foot for several minutes. I had not been able to balance on one foot for more than a second or two for at least five or six years. And, of greatest importance to me, I've cut back to only taking one 100/25 Sinemet and two 1.25 mg "dissolving" Selegeline tablets per day, and my mind is—for the first time in many years—as clear as it was before I was diagnosed with Parkinson's. I have no idea what tomorrow may bring, but right now I feel blessed.

5

LDN in Cancer

Many scientists orchestrate a dozen or more diverse research projects during their careers. Often, they're mapping out their next project and arranging funding before the current project is half complete. Ian Zagon is an exception. For more than twenty-five years he's studied the effects of opiates and opiate antagonists on cell proliferation and tissue healing.

What is it about naltrexone and related opiate antagonists that sparks such passion? Early on, it was the chameleon-like ability of opiate antagonists to evoke results that were the opposite of those expected. Like a circus animal with a penchant for finding new tricks, the opiate antagonist managed to captivate its keeper.

In 1980 Zagon set out to elucidate the cellular changes responsible for the lower birth weights seen in infants born to heroin addicts. To confirm that any observed effects were from the opiates, Zagon employed the standard approach, which is to use opiate antagonists in place of opiates in animal studies.

To his amazement, Zagon found that blocking the opiate receptor continuously with high doses of the opiate antagonist naloxone stimulated cell growth, whereas low doses of naloxone caused an intermittent receptor blockade that resulted in a rebound effect or supersensitive reaction. The result was the opposite of what was expected—the study showed that low doses of naloxone inhibited cell growth in an exaggerated fashion. This unexpected effect on cell growth was the first enticer.

Over the years, hints of the opiate antagonist's ability to heal have fascinated Zagon, spurring him on to study the effects of opiate antagonists in cell growth and wound healing. These early studies, assisted by a steady stream of technological advances, paved the way, exposing the ultimate potential of the opiate antagonist—its ability to affect cell proliferation in cancer. The research efforts of Zagon and the results of the clinical trials using opiate antagonists that he and other researchers have conducted in cancer are the focus of this chapter.

Cancer Terminology

Cancer refers to diseases characterized by the development of malignant tumors. Malignant tumors that stem from the epithelial cells that cover the surface of the skin and inner organs are called carcinomas, from the word "carcinos," a term coined by Hippocrates in about 400 B.C. Because malignant tumors have the ability to invade or spread (metastasize) into surrounding tissues and organs, in the first century, the Roman physician Aulus Cornelius Celsus chose the term *cancer*, which in Latin means crab. However, Celsus used the word *cancer* to describe malignancies confined to the skin, including wounds resulting from sepsis, gangrene, and ulcerations.[2]

A hundred years later, the physician Galen used the term *oncos* for growths or lesions that appeared cancerous. At the beginning of the nineteenth century *carcinoma* became a synonym of *cancer*. Today, the study of cancer is known as oncology, and doctors who specialize in cancer and its treatment are called oncologists. The term *oncological diseases* is the preferred term for describing conditions of cancer.[3]

Cancer Mortality Rates

Cancer mortality rates refer to the number of deaths with cancer as the underlying cause that occur in a specified population in one year. According to a 2004 report from the U.S. Department of Health and Human Services Center for Disease Control, cancer is the second leading cause of death in the United States. It's expected to be the leading cause of death within the next decade.[4]

Cancer Therapies

A number of therapies are available for patients with cancer depending on the type of cancer, its stage, and its proximity to other organs. Common therapies include surgery, radiation therapy, chemotherapeutic agents and targeted cancer therapies, which are drugs that block the growth and spread of cancer by interfering with specific molecules involved in cancer cell development and tumor growth.

Because scientists call these molecules "molecular targets," these therapies are sometimes called "molecular-targeted drugs," or something similar. By focusing on molecular and cellular changes that are specific to cancer cell growth, targeted cancer therapies may be more effective than current treatments and less likely to damage normal cells.[5] In 2005, the National Cancer

Institute (NCI) developed a translational medicine program that correlates laboratory research with clinical studies of new drugs. The use of low dose naltrexone (LDN) in cancer is considered a targeted cancer therapy.

Integrative Therapies

The National Center for Complementary and Alternative Medicine (NCCAM) division of the National Institutes of Health works together with the National Cancer Institute to study promising alternative medical treatments. The NCCAM also maintains a list of approved alternative cancer therapies and a database of clinical trial information.[6]

The Approval Process for New Cancer Therapies

Every year a number of drugs and biologics, either newly discovered compounds or ones that have been previously approved, are developed or studied for their use as cancer treatments. The U.S. Food and Drug Administration (FDA) is responsible for ensuring that these compounds are safe and effective before they're put on the market.

The FDA is also required to make promising treatments available as quickly as possible, and according to *compassionate use* policy they must make unapproved investigational drugs available before the completion of clinical trials if no approved therapies for a condition exist. A complete list of drugs, biologics, and vaccines approved as cancer treatments and that are under investigation can be found at the National Cancer Institute Website.[7]

Cancer Development

Cancers develop at the cellular level. As the building blocks of organs, cells are the body's smallest structures capable of performing all of the processes that define life such as transporting oxygen, digesting nutrients, reproducing, thinking, and moving. Cells also have a programmed death known as apoptosis.

Cell Reproduction

Worn-out, dying, or injured cells are replaced with new cells. The rate in which cells reproduce depends on various factors including disease and injury. For instance, white blood cells proliferate at a faster rate during infections, and endocrine organs respond to injury by regenerating damaged cells. Cells reproduce when they receive signals from growth factors or from contact with other cells.

Cell Cycles and Cancer

Cells reproduce through a series of steps known as growth cycles. If any of these processes goes awry, a cell may become cancerous. Cancer cells ignore signals that normally tell them to stop dividing, or to specialize, or to die and leave the circulation. Growing rapidly and unable to recognize their boundaries, cancer cells can spread to other areas of the body.

Cyclin-Dependent Kinases

Several different proteins normally control the timing of events in cell cycles. These regulations are necessary for normal cell division. The loss of this growth regulation defines cancer.

Growth cycles in cells are controlled by protein enzymes known as cyclin-dependent kinases. Each of these kinases forms a complex with a particular cyclin protein. This complex then binds and reacts with the cyclin-dependent kinase. In the process, the kinase enzymes add phosphate groups to various proteins. The phosphate molecules change the structure of the affected proteins, which can activate or inactivate the protein, depending on its function.

P53 is an important cell cycle protein with the properties of a transcription factor. Transcription is a step in cell replication. P53 binds to DNA and activates transcription of the protein p21. In a subsequent step, p21 blocks the activity of a cyclin-dependent kinase required for progression through the G1 step of the growth cycle. This block allows the cell time to repair any defective DNA before it is replicated. If the DNA damage is too severe, p53 orders the cell to commit suicide. These steps of the cell growth cycle are blocked in various cancer cell therapies, including opiate antagonists.

The p53 mutation is the most common gene mutation leading to cancer. Mutations in genes such as p53 that normally suppress tumor growth lead to cancer.

DOMINANT MUTATIONS. In cancer cells, several gene mutations occur that ultimately cause the cell to become defective. In dominant mutations one gene in the pair is abnormal. For instance, if a genetic mutation causes production of a protein that constantly activates the growth factor receptor, the cell must constantly divide. When the dominant function of the cell is affected by a gene mutation—for example, when the mutation causes the growth factor receptor to be continuously activated—the dominant gene is called an oncogene. This means it's capable of inducing cancer.

RECESSIVE MUTATIONS. In recessive mutations, both genes in the pair are damaged. For example, a normal gene called p53 produces a protein that turns off the cell cycle, thereby controlling cell growth. The p53 gene primarily functions to repair or destroy defective genes, which helps reduce cancer cell growth.

If only one p53 gene in the pair is mutated, the other gene can still control the cell cycle. If both genes are mutated, the off switch malfunctions and cell division is no longer under control.

NORMAL VS. ABNORMAL CELL DIVISION. Normal cell division requires external growth factors. When growth factor production is impaired, cells can't divide. Cancer cells don't require positive growth factors. They divide in the absence of growth factors.

When in contact with other cells (*contact inhibition*), normal cells stop dividing. That is, normal cells divide to fill in gaps in tissue. When gaps are filled, normal cells stop dividing. This property is lost in cancer cells. Cancer cells continue to grow even when large masses of cells form.

Normal cells age and die through a process of apoptosis and are then replaced by new cells in an orderly fashion. Most normal cells can divide about fifty times before they're programmed to die, which is related to their limited ability to replicate their DNA. When cells replicate, the ends or *telomeres* of their DNA shorten. In growing cells, the enzyme telomerase repairs these lost ends. In cancerous cells, telomerase is activated continuously, allowing for unlimited cell divisions.

Normal cells stop dividing and go on to die when DNA is damaged or cell division is abnormal. Cancer cells keep dividing even when the DNA is severely damaged. These progeny cancer cells contain the abnormal DNA, and as they divide they accumulate even more damaged DNA.

ABNORMAL CELL DIVISION. When active oncogenes are expressed or when genes that normally suppress tumor formation are lost, abnormal cell division can occurs. This happens: when part of a gene is lost or deleted, when part of a chromosome is rearranged and transposed to the wrong place (translocation), or when an extremely small defect occurs in the cell's DNA, causing an abnormal DNA blueprint and production of a defective protein.

Cells Gone Wild

Cancer cells behave independently. That is, cancer cells can be regarded as "cells gone wild." In their haste for uncontrollable growth, cancer cells form tumors through a series of steps. The first step is hyperplasia. Hyperplasia is characterized by an increased number of cells caused by uncontrolled cell division. The cells in hyperplasia, although increased in number, usually appear normal.

In the second step of cancer cell growth, cells appear abnormal, causing a condition of dysplasia. In the third step, cells become markedly abnormal and begin to lose their functional properties. These cells are called *anaplastic*.

In the third step, if anaplastic cells remain in their usual location the tumor is referred to as in situ. In situ cells aren't invasive and are considered potentially malignant.

In the last step of cancer cell growth, the cells in the tumor metastasize or spread out of their areas and have the ability to invade other tissues. Cancers that do not move on to other locations are considered noninvasive or benign.

Types of Tumors

Tumors are classified by the type of cells from which they arise

- Carcinomas, the most common cancers, stem from altered epithelial cells.
- Sarcomas stem from malignant white blood cells.
- Lymphomas—cancers of the B-lymphocyte white blood cells—derive from bone marrow.
- Myelomas are cancers of immunoglobulin-producing white blood cells

Angiogenesis

Tissues contain blood vessels that transport nutrients and oxygen to their cells. Both cancerous and normal tissue cells need nutrients to grow. As tumors grow and enlarge, the cells in their central area can't receive nutrients from the tissue's blood vessels. These tumors must form their own blood vessels.

Angiogenesis refers to the creation of new blood vessels. Professor Judah Folkman of Harvard Medical School found that tumors have the ability to produce new vessels. Through the process of angiogenesis, tumor cells make angiogenic growth factors that induce the formation of new capillary blood vessels. Although the cells that form these new blood vessels are inactive in normal tissue, the tumor's angiogenic growth factors activate these blood vessel cells and order them to divide. The new capillary blood vessels that grow in tumors allow for tumor growth and metastasis.

Both cells from tumor tissue and cells from the tumor's blood vessels can cross into other blood vessels and spread into lymphatic tissue and other organs. Tumor tissue cells give rise to dormant tumors that take up residence in other locations. In contrast, the angiogenic tumor cells produce new blood vessels in their new location, which causes rapid tumor growth.

Ian Zagon and Cancer Research

Ian Zagon is the director of the Program on Education in Human Structure in the Department of Neural and Behavioral Sciences at Pennsylvania State University's College of Medicine. With additional titles of Distinguished

Professor and Distinguished Educator, Zagon is also a professor in neuroscience and anatomy and a member of the Specialized Cancer Research Center.

Areas of Research

Zagon's primary areas of research include the role of opioid peptides and opioid receptors in development, cancer, cell renewal, wound healing, angiogenesis, corneal renewal, neurodegeneration, and Crohn's disease. Working with specialists in the Departments of Medicine, Health Evaluation, Surgery, Ophthalmology, and Pathology, Zagon focuses on translating scientific discoveries from the laboratory to the bedside.

Discovery of OGF

During the course of their research, Zagon and Dr. Patricia McLaughlin discovered that opioids can act as growth factors in neural and non-neural cells and tissues. Zagon found that one native opioid, methionine-5 enkephalin, exerted a negative influence on growth at low doses and a stimulatory effect when used at high doses. Zagon named this factor *opioid growth factor* (OGF) to signify its role as a growth factor, in addition to its ability to act as a neurotransmitter.[8]

Unlike chemotherapy, OGF doesn't destroy cancer or other cells. Therefore, it is not cytotoxic. However, OGF halts cell growth and is thought to allow immunological mechanisms, for instance macrophages and natural killer (NK) lymphocytes, to accomplish the task of destroying cancerous cells.

LDN vs. OGF

In his early studies of opiate antagonists on cell growth, Zagon discovered that low doses of opiate antagonists such as naloxone and naltrexone blocked opiate receptors for approximately four to six hours. This resulted in increased production of endogenous opiates once the blockade ended. In subsequent studies Zagon determined that the primary importance of this blockade is the increased production of opioid growth factor (OGF). Administering OGF has an effect similar to that caused by administering LDN. However the concentrations of OGF derived from the direct administration of OGF are much higher than the concentrations of OGF induced by LDN.

In some instances, LDN may be unable to induce sufficient OGF production. This may be due to peptide deficiencies, loss of the OGF receptor or other metabolic changes associated with cancer development. And in some cases, such as squamous cell head and neck cancers, a displacement of the OGF-

OGFr complex contributes to cancer cell development and provides a growth advantage.[9] In these circumstances and in cancers where OGF production is typically low, Zagon and his team use pure OGF or naltrexone metabolite derivatives rather than low doses of naltrexone or naloxone.

The Role of Opioid Growth Factor

Opioid growth factor protein interacts with the opioid growth factor receptor (OGFr) found on the cell nucleus. The OGFr is a nuclear-associated receptor for OGF that Zagon and his team have successfully cloned and sequenced. Bound to its receptor, OGF affects the growth and differentiation of cells and tissues. Many cancer cells have been found to have OGF receptors. In humans, the OGF receptor is highly expressed in the heart and liver, moderately expressed in skeletal muscle and kidney tissue and to a lesser extent in brain and pancreas. The OGFr is also expressed in fetal tissues including liver and kidney.[10]

When LDN or OGF is administered, OGF reacts with the OGFr on cell nuclei and forms a complex. The OGF-OGFr complex influences cell-growth pathways and arrests cell growth. The National Cancer Institute defines OGF as an endogenous pentapeptide with potential antineoplastic and anti-angiogenic activities. The NCI drug dictionary defines OGF as binding to and activating the OGFr present on some tumor cells and vascular cells, thereby inhibiting tumor cell proliferation and angiogenesis.[11]

Zagon has found that the native complex of OGF and OGFr reduces cell growth in certain types of cancer, and it contributes to the maintenance of cell replication equilibrium.[12] In summary, the complex of OGF and OGFr serves as a tonically active system that maintains cellular homeostasis and targets the cyclin-dependent inhibitory kinase pathway.

OGF's Effect on Cell Pathways

OGF inhibits cell growth in cancer by targeting specific cyclin-dependent inhibitory kinase pathways. In their research laboratory at Penn State, Zagon and his colleagues have demonstrated that the OGF-OGF receptor axis uses the p16 pathway to inhibit head and neck cancer cell growth.[13] Shortly after making this discovery, Zagon and his team determined that the OGF-OGF receptor axis uses the p21 pathway to inhibit pancreatic cell cancer growth.[14]

Effects of OGF on Cancer Cells

In the December 2007 issue of *Neuropeptides*, Zagon and his team reported the results of a study designed to determine the role of OGF and naltrexone on the migration, chemotaxis, invasion, and adhesion of human cancer cells. The study involved cultured human pancreatic and colon cancer cells and cells derived from squamous cell carcinoma of the head and neck.

Using high concentrations of naltrexone and related opiate antagonists to stimulate cell growth and low concentrations to inhibit cell growth, Dr. Zagon and his team showed that these effects are independent of cell migration, chemotaxis, invasion, and adhesive properties. Furthermore, neither naltrexone nor OGF affected these biological properties of cancer cells. Zagon also tested a variety of endogenous and exogenous opioid compounds and showed that these compounds also had no effect of the biologic properties of cancer cells.[15]

Modulation of Angiogenesis

Zagon and his team have also determined, in animal studies using chick eggs, that opioid growth factor has a significant inhibitory effect on angiogenesis. Their studies showed that both the number of blood vessels and the blood vessel length were decreased in vivo. The study concluded that opioid factor is a tonically active peptide with a receptor-mediated action in regulating angiogenesis in developing endothelial and mesenchymal vascular cells.[16]

Pancreatic Cancer

Pancreatic cancer is the fourth-leading cause of cancer mortality in the United States. Zagon and his team have been studying the effects of OGF in pancreatic cancer for the past decade. In their laboratory, they've found that OGF controls cell growth in pancreatic cancer. Pancreatic cancer cells have OGF receptors that, when bound with OGF, inhibit additional cancer cell growth. Because cancer cell growth is unregulated and cancer cells grow so quickly, the body can't produce enough OGF on its own to bind to all the OGF receptors. When OGF binds to the OGF receptors on cancer cells, cancer cell proliferation is inhibited. Increasing OGF levels with LDN or OGF accelerates this process.

In a study published in the April 2007 *International Journal of Oncology*, Zagon and his team described a study in which they demonstrated that overexpression of the opioid growth factor receptor induced by low dose naltrexone also enhances growth inhibition in pancreatic cancer.[17] That is, LDN, by increasing both OGF and the number and density of OGF receptors, increases the number of OGF-OGFr complexes available to inhibit cancer cell growth.

Phase 1 Trials

Phase 1 trials of OGF conducted in Zagon's laboratory using human pancreatic cancer cells and xenografts of nude mice demonstrated that opioid growth factor inhibited pancreatic cell growth. In addition, in this trial, which was funded by the National Institutes of Health, the safety of OGF and its maximum tolerated dose based on hypotension were determined in twenty-one patients with unresectable pancreatic adenocarcinoma. In this part of the trial, OGF was administered either subcutaneously or intravenously. During the intravenous phase of the study, two patients had resolution of associated liver metastases, and one patient showed regression of the pancreatic tumor.[18]

INNO-105

OGF is a natural-occurring molecule also known as (Met-5)-enkephalin with potential as an anticancer compound. Following the Phase 1 trial at Penn State University, Innovive Pharmaceuticals, a privately held biopharmaceutical company headquartered in New York, licensed OGF as an anticancer agent with the code name INNO-105. In November 2005 Innovive began its own Phase 1 clinical trials on a variety of different cancers but discontinued development of INNO-105 in late September 2006.[19]

Phase 2 Trials and Combination Therapies

In February 2008, Zagon's team was still recruiting patients for phase 2 clinical trials of OGF in people with advanced pancreatic cancer that cannot be removed by surgery.[20] Zagon's team has also conducted a trial of combination chemotherapy with gemcitabine and OGF. Gemcitabine exerts its effects through inhibition of DNA synthesis and has the drawback of limited survival benefits. Preclinical evidence from this trial showed that OGF enhances growth inhibition when used together with gemcitabine, which is the standard of care for advanced pancreatic neoplasia.[21]

Lipoic Acid/OGF Protocol

Dr. Burton Berkson of Las Cruces, New Mexico, describes the long-term survival of a patient with pancreatic cancer treated with a combination of intravenous alpha-lipoic acid and low dose naltrexone. In this report, Berkson indicates that although the pancreatic tumor did not decrease in size it showed no disease progression, and the patient reported having an improved quality of life.[22]

Head and Neck Cancer

Head and neck squamous cell carcinoma represents 5.5 percent of malignancies worldwide. Approximately 30,000 new cases and 11,000 deaths are reported in the United States annually. Studies performed by Zagon and his team show that the complex formed by the OGF-OGF receptor inhibits cell growth in these cancers, and that deficiencies of the OGF receptor in head and neck squamous cell carcinoma are responsible for its rapid rate of tumor progression.[23] Related studies at Penn State using OGF alone and in combination with paclitaxel resulted in a significant reduction in tumor weight and increased survival.[24]

Lymphoma

In a report presented at the April 2007 conference on opiate antagonists, Berkson presented a report of a case of follicular lymphoma he reversed in a sixty-one-year-old male patient using LDN and nine intravenous treatments with alpha-lipoic acid. The patient refused standard medical treatments and declined further treatment with alpha-lipoic acid but continued on a six-month course of naltrexone alone. Follow-up studies showed improvement up to one year following the end of treatment.[25]

In his report Berkson described using the protocol for LDN taken at night originally recommended by Bernard Bihari in his anecdotal accounts, described on his Web site, of successfully treating several patients with lymphoma.

Restoring Homeostasis

When asked what he considered the common link in all of the disorders that respond to LDN, Zagon stated that these are all disorders that benefit from LDN's influences on cell proliferation, for instance tumor cell growth. Zagon also stated that LDN restores homeostasis. That is, it allows the body to heal itself.[26]

Dr. Nicholas Plotnikoff, a professor in the College of Pharmacy in the College of Medicine at the University of Illinois at Chicago describes metenkephalins as having immunological properties similar to those of the cytokines interleukin-2 (IL-2) and interferon-gamma (IFN) in that metenkephalins have potent antiviral and antitumor properties. Specifically, Dr. Plotnikoff reports that metenkephalins increase levels of CD8 T lymphocytes, CD4 T lymphocytes, IL-2, and natural killer (NK) lymphocytes. In addition, metenkephalins have been shown to cause increased blastogenic responses to mitogens.[27] These properties allow the body to fight cancers, viral infections, and inflammatory disorders, thereby helping the body heal itself.

Neuroblastoma

In an animal study published in 1983, Zagon and McLaughlin demonstrated that naltrexone modulates the tumor response in neuroblastoma-inoculated mice.[28] In a series of studies conducted in the late 1980s, Zagon confirmed the presence of opioid receptors on neuroblastoma cells and demonstrated growth inhibition of neuroblastoma in mice with low dose naltrexone. In this study, Zagon found that very small doses of naltrexone (0.1mg/kg/day) inhibited tumor growth, prolonged survival in the mice that developed tumors, and protected some mice from developing tumors altogether.

In a study published in 2005, Zagon and McLaughlin demonstrated significant tumor cell growth inhibition by OGF in human cell cultures of pancreatic cancer, head and neck cancer, and neuroblastoma. In each experiment, metenkephalin acted as a negative regulator of tumor development and significantly suppressed tumor appearance and growth.[29]

Colon Cancers

In animals studies of mice, including nude mice inoculated with human colon cancer cells, OGF was shown to inhibit tumor cell growth, and it delayed and prevented growth of human colon cancer xenografts.[30] In the nude mice study published in 1996, mice were administered daily doses of 0.5, 5, or 25 mg/kg opioid growth factor, [Met5]enkephalin. More than 80 percent of the mice receiving opioid growth factor beginning at the time of tumor cell inoculation did not exhibit neoplasias within three weeks, in comparison with a tumor incidence of 93 percent in control subjects.

Seven weeks after cancer inoculation, 57 percent of the mice given OGF did not display a tumor. OGF delayed tumor appearance and growth in animals developing colon cancer with respect to the control group. The suppressive effects of OGF on oncogenicity were opioid receptor mediated. In addition, OGF and its receptor were detected in transplanted HT-29 colon tumors. Surgical specimens of human colon cancers also showed evidence of OGF. The study concluded that a naturally occurring opioid peptide acts as a potent negative regulator of human gastrointestinal cancer and may suggest pathways for tumor etiology, progression, treatment and prophylaxis.

Metastatic Solid Tumors

A team of researchers located in Milan, Novosibirsk, Tel-Aviv, and Locarno, have applied principles rooted in psychoneuroimmunology in a collaborative study in which the researchers treated metastatic solid tumors with melatonin, with and without naltrexone, in combination with interleukin-2.

The rationale for the study was that naltrexone and melatonin, by activating Th1 lymphocytes and suppressing Th2 lymphocytes, should enhance the increase in total lymphocyte count induced by Il-2. This preliminary study showed that the addition of naltrexone further amplifies the absolute lymphocyte count, which, in turn, enhances the anticancer efficacy of IL-2. The increase in lymphocyte count attributed to the addition of naltrexone was significantly higher than that achieved with melatonin and IL-2 alone. In contrast, patients treated with IL-2 and melatonin without naltrexone showed primarily an increase in eosinophils.[31]

Renal Cancer

Working with a team of oncologists and urologists at Penn State, Ian Zagon has demonstrated that OGF interacts with the OGF receptor to directly inhibit proliferation of renal cell carcinoma in tissue culture. In this study human renal cancer cells (CAKI-2) were grown using routine tissue culture techniques. A variety of natural and synthetic opioids including OGF, opioid antagonists, and opioid antibodies were added to renal cancer cell cultures to determine the role of these peptides in renal cell carcinoma.

The experiments were repeated in serum-free media and with four other renal cancer cell lines. Immunocytochemistry methods were used to examine the presence of OGF and its receptor. The study results demonstrated that OGF was the most potent opioid peptide to influence human renal cell carcinoma. OGF depressed growth within twelve hours of treatment, with cell numbers reduced by up to 48 percent of control levels.

In addition, OGF action was shown to be receptor mediated, reversible, not cytoxic, neutralized by antibodies to the peptide, and detected in the human renal cell carcinoma lines examined. OGF appeared to be autocrine produced and secreted, and was constitutively expressed. The researchers concluded that OGF tonically inhibits renal cancer cell proliferation in tissue culture, and may play a role in the pathogenesis and management of human renal cell cancer.[32]

Other Cancers

According to a report by the MedInsight Research Institute, a nonprofit research institution in Baltimore, the following cancers have either been shown to have OGF receptors and/or have been anecdotally reported to respond to OGF and/or OGF-boosting mechanisms such as LDN: breast cancer, cervical cancer, colon and rectal cancer, gastric cancer, glioblastoma, head and neck cancer, Kaposi's sarcoma, lymphocytic leukemia, liver cancer, B cell lymphoma, T cell lymphoma, malignant melanoma, neuroblastoma, ovarian cancer, pancreatic cancer, prostate cancer, renal cell carcinoma, small cell and non-small cell lung cancer, throat cancer, tongue cancer, and uterine cancer.[33]

Each week, Ian Zagon hears from physicians, including veterinarians, and patients from all over the world. They seek advice regarding the use of LDN and OGF or they write to express their gratitude as they describe their—or their patient's—success in using LDN. Zagon has shared some of this correspondence in confidence. He admits that these anecdotal patient accounts are the most gratifying aspect of his work.

Case History of LDN and OGF in Lung Cancer

On November 20, 2007, Dr. Winston Shaer, a plastic surgeon from Capetown, South Africa, contacted Zagon for advise regarding his wife Barbara, who had recently been diagnosed with stage 4 non-squamous, non-small cell lung cancer (NSCLC). Barbara was found (via transbronchial needle biopsy and PET scan) to have a confluent mass of tumor and nodes surrounding the right main bronchus, with partial obstruction of the right upper lobe bronchus. In addition cancerous nodes were detected at the carina and anterior mediastinum on the right as well as some nodes to the left of the trachea and two small nodes at the porta hepatis. No bone or brain metastases were seen. At the time of her diagnosis, Barbara was described as an attractive, intelligent, healthy and fit sixty-year-old Type 1 patient.

At this time, Barbara had been on chemotherapy for one week with no ill effects. Her oncologists planned a twelve-week course of Cisplatin/Gemsa plus Avastin with four cycles of alternating courses with and without Avastin. A PET scan was scheduled to be performed after the third cycle of chemotherapy to assess the mediastinum and determine if radiation was needed.

In addition to this protocol, Dr. Shaer had recommended that his wife follow a primarily vegetarian diet, with the exception of a 100-gm fillet at night, an occasional egg, and goat cheese. He was also administering 25 grams of intravenous vitamin C on alternate days and recommended that Barbara take the supplements Agaricus, Metagenics whey in a strawberry and blueberry shake, omega-3 oils, and Moducare, a natural plant-based immunomodulator. In a November 2007 e-mail to Zagon, Shaer inquired if low dose naltrexone might help. In response, Zagon suggested trying low dose naltrexone at a dose of about 4.5 mg daily as long as Barbara was not taking any opiate-based medications and that her condition was closely monitored.

Zagon had previously demonstrated that both LDN and OGF can be used to enhance the effects of chemotherapeutic agents. However, Zagon cautioned that LDN and OGF should not be used simultaneously since LDN works to increase OGF levels. Zagon also recommended that Shaer consider treatment with opiod growth factor instead of LDN and gave Shaer contact information for Moshe Rogosnitzky in Israel (see Resources for more information).

On January 10, 2008, Shaer again wrote to Zagon and reported that Barbara was very ill although she had been using LDN and had completed her third cycle of chemotherapy. She was scheduled to have a CAT/PET scan to gauge her response to chemotherapy. Shaer asked if Barbara might use opioid growth factor in place of LDN administered intravenously with the vitamin C. He also indicated that he had been giving her vitamin C injections three times weekly and that his wife was getting weary of needles and pain. Zagon replied that OGF would be a good option.

On February 15, 2008, Shaer wrote Zagon and reported that Barbara had had a PET scan two days before which demonstrated " a huge improvement and an almost complete response to treatment. She has a residual 10 mm nodule at the right apex, a 10 × 15 mm nodule at the helium, and a tiny, adjacent, barely discernible mediastinal area of activity. The lung nodules both demonstrate a 95 percent reduction in activity with no distant or further evidence of metastasis. The lung fields are otherwise perfect."

At this time Barbara had completed four cycles of Cisplat/Gemza/Avastin along with two additional cycles of this protocol. She'd had controllable moderate hypertension, no weight loss, and felt extremely well. Along with LDN and the original complementary regime, Shaer had added within the month bovine lactoferrin, inositol hexaphosphate, alpha lipoic acid, L-glutamine, melatonin and glutoximine.

In March 2008 Shaer wrote to Zagon and reported that his wife was "vastly improved" and was scheduled to have her final dose of chemotherapy within the week. She was scheduled to have a repeat PET scan six weeks later to allow any residual tumor, if present, time away from the cytoxic induced resting phase. In other words if the tumor was still present it would likely show up by this time.

Shaer wrote that he thinks LDN is a huge and very critical component of the treatment protocol he helped devise with the assistance of Ian Zagon and Moshe Rogosnitzky. However, because a number of complementary agents, chemotherapeutic agents and dietary interventions were also used, he explained that it's impossible to determine definitively what role LDN played.[33]

Although Shaer's wife Barbara used a number of different therapies including LDN, it isn't unusual for patients who seek out LDN to also use one or more alternative or complementary medical therapies. For this reason, anecdotes describing success with LDN are limited because other factors are often involved. This story is important, though, because it illustrates how patients and physicians around the world are including LDN in their treatment protocols.

Dr. Bihari's Cancer Patients

On his Web site, Bernard Bihari reports that as of March 2004 he had used LDN to treat as many as 450 patients with a variety of different cancers, including breast, lung, liver, ovarian, prostate, pancreatic, and renal cancers as well as various leukemias and lymphomas, neuroblastoma and multiple myeloma. Bihari attributes the improvement he has seen in these cancers to increased levels of metenkephalins, increased number of opiate receptors on cancer cell membranes, and increased levels of CD8 T lymphocytes.[34]

Bihari's Cancer Records

Because of Bihari's reported anecdotal success in treating cancer, the National Center for Complementary and Alternative Medicine (NCCAM) Division of the National Institutes of Health identified LDN as a potential alternative cancer therapy. To investigate these claims, they requested that the Agency for Healthcare Research and Quality (AHRQ), which develops scientific information for other agencies based on evidence-based medicine and other performance measures, study Bihari's patient records.

In perusing these records the AHRQ investigators were required to determine which cases fit the criteria for best-case examples. To be included as best cases, the records required documentation of the diagnosis of cancer with records of pathology studies that also evaluated the appropriate tumor endpoint so that comparison could be made. In addition the patients could not be using any other cancer treatments and records of previous cancer treatments had to be documented. The sites of the cancer had to be documented with histological studies and dates of recurrence or metastasis noted. The patient's general health and the description of the LDN treatment needed to be documented.

After promising cases were identified the researchers had to contact the patients and obtain permission to abstract their files. Patients, or their next of kin if deceased, were interviewed by phone. Case reports were then compiled and sent to the Southern California Evidence-Based Practice Center to see if they met the National Cancer Institute's criteria for study inclusion. Of the twenty-one cases that the researchers reviewed, only three cases met the NCI criteria. Consequently, the AHRQ determined that the naltrexone cases provided insufficient evidence to determine the likely benefits of LDN at that time.[35]

It's important to note that Bihari treated many patients with advanced cancer who had already exhausted all other available treatment options when they consulted him. It is unfortunate but understandable that some of the records he maintained were incomplete. With the resources available in a formal clinical trial, Bihari's records may have told a different story.

The Future of LDN and OGF in Cancer

Many of the latest therapies under development for cancer are designed to target genetic and metabolic factors that are essential for the development, growth, and proliferation of malignant cells. The role of molecular genetics and proteomics and the application of molecular technology in assessing cancer cell metabolism are the tools that drive these new therapies. Experts in the field caution that the genetic transformations and proteomic alterations achieved by these therapies will only have relevance to disease if they do not result in altered and impaired cellular and metabolic function in other cells.[36]

Therapies such as OGF and LDN target cancer cell metabolism without disturbing normal cells, making them potentially desirable therapies. But the benefits of these compounds in cancer don't stop here. Dr. Zagon is conducting studies and has already published several papers describing the overexpression of the OGF receptor in some cancer cells, which naturally inhibits tumor growth. Gene therapy used to increase the OGF receptor on tumor cells is another potential cancer therapy that would enhance the benefits of both LDN and OGF.

Alternately, Zagon reports that knocking down the OGF receptor gene with either antisense OGF receptor cDNA or siRNA directed to the OGF receptor markedly enhances cell growth, with potential applications for wound healing.[37]

6

LDN in Autism Spectrum Disorders

The ability of low dose naltrexone (LDN) to halt disease in a wide range of unrelated conditions can make favorable claims about this drug seem exaggerated. This is particularly true in autism spectrum disorder, a condition that includes a broad range of related disorders with symptoms ranging from mild disturbances of social interaction to complete social withdrawal.

This ability to offer benefits in many diverse disorders is one of the most common criticisms of LDN mentioned by its critics.[1] We asked the doctors we consulted for this book what they considered the common link in illnesses reported to respond to naltrexone. They stated that the conditions with the potential to improve with LDN are conditions that benefit from alterations in cell proliferation or from a reduction in chronic inflammation.

Simply put, LDN is believed to restore homeostasis. By invoking changes in immune function, LDN helps the body heal itself. This chapter focuses on the effects and outcomes associated with naltrexone therapy in autism spectrum disorder.

Autism Spectrum Disorder

The National Institute of Neurological Disorders and Stroke (NINDS) division of the National Institutes of Health (NIH) defines autism spectrum disorder (ASD) as a family of conditions characterized by impaired social interaction and problems with verbal and nonverbal communication. Autism spectrum disorders include classical autism, which is the most common of the autism spectrum disorders; Asperger syndrome; Rett syndrome; childhood disintegrative disorder; and pervasive developmental disorder not otherwise specified (PDD-NOS). Symptoms in autism spectrum disorder can range from mild to disabling.

The NINDS fact sheet on ASD also describes children with autism as having unusual, repetitive, or severely limited activities and interests and difficulty interpreting what others are thinking or feeling because they're unable to understand social cues, such as tone of voice or facial expressions. Also, children with autism are reported to avoid eye contact and avoid watching other people's faces for clues about appropriate behavior. Although many children with autism are reported to have a reduced sensitivity to pain and rarely cry, they may be abnormally sensitive to sound, touch, or other sensory stimulation. These unusual reactions may contribute to behavioral symptoms such as a resistance to being cuddled or hugged.

Children with autism also appear to have an increased risk for certain coexisting conditions, including fragile X syndrome (which causes mental retardation), tuberous sclerosis (in which tumors grow on the brain), epileptic seizures, Tourette syndrome, learning disabilities, and attention deficit disorder. For reasons that are still unclear, about 20 to 30 percent of children with autism develop epilepsy by the time they reach adulthood. Experts estimate that three to six children out of every 1,000 will have autism or related disorders classified as autism spectrum disorders. Males are four times more likely to develop autism than females.[2]

Opiates, Endorphins and Autism

The psychologist Jaak Panksepp from Bowling Green State University in Bowling Green, Ohio, one of the leading theorists and researchers in biochemistry and autism, reports that the discovery of the opiate receptor in 1972 provided a better understanding of autism. Early animal studies show that opiates increase the tendency for social isolation, whereas when animals were given opiate antagonists they became more disturbed by social isolation and became eager for gentle and friendly social contact.

These observations suggested to Dr. Panksepp that disturbances in endorphins might be responsible for autism. Supporting this theory, several studies found that some autistic children had high levels of endogenous opioid compounds. However, early studies found no consistent correlation between autism and beta endorphins. Panksepp, in an interview with Stephen Edelson in 1997, explains that the variable amounts of endogenous opiates in children with autism may account for the reason that some children with autism respond to naltrexone and others do not.[3]

The Opiate Hypothesis in Autism

Panksepp is credited with the opioid hypothesis, which suggests that childhood autism may result from excessive brain opioid activity during the neonatal period. Strong support for this hypothesis focuses on similarity between symptoms in autism and those seen in young animals injected with endogenous opioids, direct biochemical evidence of abnormalities in levels of endogenous opioids in autism, and effectiveness of naltrexone in studies of autism.[4]

Gluten, Casein, and Endorphins

Gluten is a protein found in wheat, rye, and barley. Casein is a protein found in milk products. Children with autism have been found to have inflammatory reactions in the gut after eating these products.

In 1991, the physician Kalle Reichelt hypothesized that gluten and casein may have an opiate effect in children with autism. Studies confirmed that in some children with autism, enzyme deficiencies resulted in poor casein digestion and excess levels of the opioid peptide casomorphine whereas gluten ingestion resulted in excess levels of the opioid peptides gluten exorphines and gliadorphins. Reichelt described long-term exposure to excess opioids contributing to the social isolation characteristic of autism.[5]

The gluten-free, casein-free diet has been widely accepted and recommended by the Autism Institute. However, not all children with autism show improvement following this diet. A long-term double-blind clinical trial sponsored by the National Institute of Mental Health is currently in progress and scheduled for completion in April 2008; preliminary results are expected to be available at the end of 2008.

LDN in Autism

Because high levels of endogenous opiates are seen in some children with autism, a number of researchers have studied naltrexone as a therapy for autism. In one of the earliest studies, French researchers using 1mg/kg/day of naltrexone observed an immediate reduction of hyperactivity, self-injurious behaviors and aggressiveness, while attention improved. In addition, social behaviors, smiling, social-seeking behaviors, and play interactions increased. In a cross-over double-blind study, three doses of naltrexone, 0.5, 1, and 2 mg/kg/day were used and the effects compared.[6]

These researchers later conducted a double-blind study using the lowest effective dose of naltrexone observed in their earlier study, 0.5 mg/kg/day, and contrasted it with a placebo. This study included clinical and biochemical evaluations. Using the C-terminal antibody assay, all of the children with autism showed marked elevations of beta endorphins at the onset of the study. In addition, 70 percent of the children exhibited abnormally low levels of adrenocorticotrophic hormone (ACTH); 60 percent showed elevations in norephinephrine; 50 percent showed elevations of arginine-vasopressin, and 20 percent showed elevations in serotonin.

Children in both groups (placebo and low dose naltrexone) showed modest clinical benefits, although marginally better overall results were seen in the

LDN group. The children with the fewest hormone abnormalities at the onset of the study showed the best response to LDN. These results also suggest a possible linkage between abnormal hormones, particularly ACTH, and the development of autism.[7]

Several other small studies of low to moderate doses of naltrexone were conducted in the early 1990s with mixed results. The largest of these trials showed that naltrexone caused a reduction in hyperactivity in children with autism.[8] However, with the development of new medications such as Paxil and Prozac, which are used to reduce symptoms in autism, interest in naltrexone for autism began to falter in the late 1990s until it was resurrected by the psychiatrist Jaquelyn McCandless.

Autism and Autoimmunity

In her book, *Children with Starving Brains*, McCandless explains that subgroups of autism exist. While some children with autism have an overactive immune system, others have suboptimal immune regulation. Because of the prevalence of brain antibodies in autism, McCandless classifies autism as an autoimmune disorder. Many other autism experts agree.[9]

Researchers at the University of California Los Angeles have discovered that children with autism and adults with autoimmune disease have high levels of autoantibodies against three peptide enzymes, gliadin, and heat shock proteins. The results of this study suggest that dysfunctional membrane peptidases and autoantibody production may result in neuroimmune dysregulation and autoimmunity.[10]

McCandless describes LDN as blocking dietary opioids, reducing autoantibody production by shifting T cell production and restoring immune balance, and causing an upregulation of endogenous endorphins that can help improve social interactions. McCandless emphasizes that social interactions may improve if positive social reactions are reinforced. In addition, McCandless notes that some children with autism who never seem to get sick develop viral infections shortly after starting LDN. She attributes this to a restoration in normal immune regulation, which causes latent infections to emerge.

Interleukin 10 in Autism

Dr. Cynthia Molloy and her colleagues at the Center for Epidemiology and Biostatistics at Cincinnati Children's Hospital Medical Center, have identified a potential mechanism for the immune dysregulation noted in autism spectrum disorder. The anti-inflammatory cytokine interleukin-4 (IL-4), which is produced by Th2 lymphocytes, is responsible for producing the regulatory cytokine interleukin-10 (IL-10). Children with autism frequently have imbalances of Th1 and Th2 cells and low levels of IL-10. Th1 cells are typically increased in autoimmune disease and Th2 cells are increased in people with allergies.

Molloy's study showed that children with autism frequently have elevations of both Th1 and Th2 lymphocytes. Therefore, both Th1 and Th2 cytokines are more highly activated even in the resting state. This causes an increased hypersensitivity to environmental stimuli. Comparing children with autism to a control group, the children with autism were found to have lower IL-10 levels although having elevations of Th1 and Th2 cells suggested that they would have elevated IL-10 levels. Molloy hypothesizes that children with autism may not be able to downregulate their Th1 and Th2 systems either because of a dysfunction in the production of IL-10 or because of a dysfunction with IL-10 activity.[11]

In a related sixteen-week, non-placebo, noncontrolled study, McCandless found that LDN caused an elevation in IL-10 levels in some children and decreased IL-10 levels in others. Overall, levels of IL-10 in the subject studies were low, confirming Molloy's results. However, the effects of gluten-free and casein-free diets and mucosal activity on IL-10 levels would also need to be taken into consideration in assessing changes.

Researchers at the Department of Pediatrics at the New Jersey Medical School in Newark have previously shown that IL-10 levels are decreased in autistic children, with gastrointestinal symptoms related to ingestion of dietary proteins but not in autistic children without hypersensitivity to dietary proteins.[12]

Developmental Growth

Ian Zagon, in a patent application describing growth regulation and related applications of opioid antagonists, explains that upon diagnosis of uterine growth retardation syndrome or any related fetal growth disorder, a regime of naloxone, naltrexone or any of their effective analogs, derivatives, or metabolites, can be developed in an effort to accelerate fetal growth to its normal state. A growth regime can be tailored to accompany diagnosis by sonography, amniocentesis and other diagnostic techniques.[13]

Safety and Efficacy of Naltrexone in Autism

Researchers at St. John's University in Queens, New York performed a literature search using Medline (1966–May 18, 2006) and the International Pharmaceutical Abstracts (1971–May 18, 2006) databases to review the efficacy and safety of naltrexone in pediatric patients with autism. Three case reports, eight case series, and fourteen clinical studies were identified as pertinent.

A review of the data showed that naltrexone has been used most commonly at doses ranging from 0.5 to 2 mg/kg/day and has been found to be predominantly effective in decreasing self-injurious behavior. The studies also indicated that naltrexone may reduce symptoms of hyperactivity, agitation, irritability, temper tantrums, social withdrawal, and stereotyped behaviors. Patients may also exhibit improved attention and eye contact.

Side effects were rare, with transient sedation noted as the most commonly reported adverse event. Small sample size, short duration, and inconsistent evaluative methods characterize the available research. The St. John's study concluded that children with autism may benefit from a trial of naltrexone therapy, particularly if the child exhibits self-injurious behavior and other attempted therapies have failed.[14]

Self-Injurious Behaviors

Low and moderate doses of naltrexone have been studied and shown to offer benefits in eating disorders, self-injurious behaviors, tics, autism, obsessive compulsive disorders, schizophrenia and post-traumatic stress disorders. In one study, the lead investigator wrote that that opiate receptor antagonists used alone are not believed to have a decisive global therapeutic effect for these conditions, but have therapeutic value when used in conjunction with cognitive and behavioral therapies.[15]

Endogenous Opiates and Self-Injury

In a study of self-injurious behavior from Geneva, Switzerland, Dr. Annabel McQuillain explains the role of endogenous opioids in self-injurious behaviors in patients with borderline personality disorders. In borderline personality disorder a dysfunctional stress response is hypothesized to underlie the acute subjective distress and the accompanying somatic symptoms that occur prior to deliberate self-harm. Soft-tissue injury causes a peripheral afferent stimulation related to neuroimmune mechanisms that are hypothesized to relieve the subjective tension by reestablishing homeostasis.

Deliberate self-harm, as distinct from suicide attempts, is seen in several psychiatric conditions, including autism, and is also one of the diagnostic criteria of borderline personality disorder. In her clinical observations of more than 200 patients, McQuillan finds that prior to self-harm, patients report subjective feelings of general distress, for example a chronically stressful personal situation or a depressive episode. An environmental trigger, usually separation, produces a characteristic pattern of intense subjective distress accompanied by dissociative symptoms. Separation distress is powerfully inhibited by beta endorphin, prolactin and to a lesser extent, oxytocin. These circuits are all developmentally related to pain pathways.

The Role of Opiate Antagonists

Opiate antagonists increase social behavior in humans and nonhuman primates, whereas the administration of opiates reduces contact seeking. Humans develop marked separation distress at around six months of age and this normally last for many years.

The endogenous opiate system participates in many physiological functions, including the physiological aspects of dependence, reward, analgesia, learning, stress and temperature regulation and neuroendocrine function. The opioid peptides are known to produce transient stress-related analgesic states and they have also been implicated in the genesis of dissociative states. Opiate antagonists cause an elevation of beta endorphins and a reduction in deliberate self-harm. The mechanism leading to self-harm, which may be related to genetic changes, imbalances of corticotropin releasing hormone (CRH), adrenocorticotropic hormone (ACTH) or endogenous opioid systems, becomes less active after sexual maturation, particularly in males. Deliberate self-harm in the form of tissue damage restores homeostasis by activating nociceptive and non-nociceptive neuro-humoral systems. Opiate antagonists restore homeostasis and reduce self-harm through similar mechanisms.[16]

7

LDN in Wound Healing and Infections

In his early studies of naltrexone, Ian Zagon discovered that high doses of naltrexone (HDN) stimulate cell growth, whereas low doses of naltrexone (LDN) inhibit cell growth. Zagon describes a patient in remission for breast cancer who experienced a recurrence of cancer after using extra doses of LDN. She mistakenly thought that because LDN worked so well for her arthritis, more would be better.

High dose naltrexone poses other risks. Because high dose naltrexone blocks the opiate receptors continuously, patients who simultaneously take heroin or other opiates (which are contraindicated) would not feel any effects of this drug. This could cause them to take additional doses, which could potentially lead to overdoses. HDN also blocks endogenous opiates from reacting with the opiate receptor. Therefore, the "feel good" effects related to endogenous opioids produced during exercise, lovemaking, or massage wouldn't be experienced. This could lead to depression and poor decisions.

An understanding of the differences in cell proliferation that occur depending on whether low or high doses of naltrexone are used can be best demonstrated by the effects of naltrexone in wound healing and bacterial infections. These topics are the focus of this chapter.

High Dose Naltrexone
vs. Low Dose Naltrexone

From his early studies of infants born to heroin addicts, Zagon and his colleagues discovered that high doses of opiate antagonists and many low doses of opiate antagonists stimulate cell growth, and low doses of naltrexone inhibit cell growth. High doses of naltrexone cause a sustained blockade of the opiate receptor, whereas low doses cause an intermittent blockade that results in increased production of endogenous opiates and their receptors. The increase in complexes formed by opioid growth factor (OGF) and its receptor (OGFr) halt cell growth.

Wound Healing

For this reason, high doses of naltrexone are very effective in conditions that benefit from increased cell growth, such as wound healing. Zagon explains that in diabetics in particular, wound healing is very slow because diabetics naturally have high levels of opioid growth factor.[1] High levels of opioid growth factor inhibit cell growth, thereby slowing the healing process. Blocking the opioid growth factor receptor continuously with high doses of naltrexone promotes wound healing. To reduce the systemic effects of HDN, topical creams can be used.

Wound healing can be enhanced dramatically by accelerating the growth and cell division (mitosis) of wounded tissue and fibroblast cells, which increases collagen production and promotes healing granulation. Fiber production can be facilitated in culture or in vivo at increased rates using high dose naltrexone. High doses of naltrexone can be subsequently or concurrently applied to wounded areas by layering or other techniques, including implanting subcutaneous pumps, or using beads or topical preparations. Another potential application is skin grafting used in the treatment of burn victims.[2]

Corneal Reepitheliazation

Patients with diabetes have an increased risk for developing corneal disorders, often as a result of surgical and non-surgical trauma. Ocular disorders secondary to diabetes are a leading cause of blindness in the western world, and corneal disorders are becoming increasingly recognized as a cause of patient morbidity related to diabetes. Ian Zagon and his team at Penn State have found that maintaining blood glucose concentrations close to the normal range helps with the poor corneal wound healing typically seen in diabetes, and improves corneal reepitheliazation, the process in which the surface epithelial cells of the cornea regenerate.[3]

Naltrexone in Corneal Disorders

Ulcers and erosions of the corneal epithelium, as well as delays in resurfacing of the cornea after wounding, are major causes of ocular morbidity and visual loss in patients with diabetes. Zagon and his team have discovered

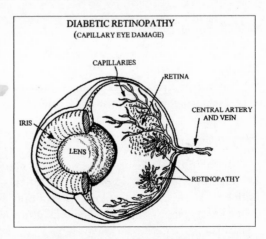

Diabetic retinopathy (Marvin G. Miller)

in animal studies of rats that topical applications of high dose naltrexone (sustained opiate receptor blockade) are effective in rescuing the dysfunctional corneal reepitheliazation in diabetics and repairing the corneal defect.[4]

Ian Zagon and his team have confirmed that topically applied naltrexone in the form of eye drops has no overt toxicity for the parameters studied. This indicates that efficacy studies for the use of naltrexone in corneal wound healing are warranted.[5]

Infections

By increasing metenkephalin levels, low dose naltrexone induces biochemical changes that reduce sepsis and interfere with viral replication. In some of Bernard Bihari's earliest studies he applied these principles to effectively use LDN as a treatment for herpes.[6] One of Bihari's first patent applications was for the use of low dose naltrexone in herpes infection. In the mid–1980s Bihari also worked closely with Nicholas Plotnikoff, an expert in the use of metenkephalins, to reduce the incidence of opportunistic infections in HIV-infected patients with AIDS-related complex (ARC).[7] The term ARC was used in the 1980s to describe patients who had certain signs of HIV infection but had not yet developed full-blown AIDS. Today, these patients are described as having HIV infection.

In a related study, researchers at the University of South Florida College of Medicine have shown that combination therapy using an anti-retroviral drug along with metenkephalin caused an increase in type 1 cytokines, primarily interferon-gamma, that help reduce the progression of HIV infection to AIDS.[8]

Viral Infections

Nicholas Plotnikoff explains that LDN's effects are primarily related to increased production of metenkephalins. He reports that the antiviral properties of metenkephalin make it an effective therapy against herpes, HIV infection, cytomegalovirus, coronavirus, influenza A, and Japanese encephalitis.[9]

Hepatitis

Bernard Bihari reports treating more than sixty patients with hepatitis C successfully with LDN. Some of these patients were treated in combination with anti-retroviral medications. In addition, he has used a combination of LDN and hypericin, the active ingredient in the herb St. John's wart, to treat hepatitis B.[10]

LDN in Lyme Disease

Lyme disease is caused by a bacterial infection transmitted by tick bites. In the United States, the usual infectious agent is *Borrelia burgdorferi* and the most common vector is the deer tick, although other ticks have been implicated. Flu-like symptoms, musculoskeletal pain, bull's eye rash, arthritic pain, psychiatric manifestations, and cardiac disturbances typically occur in Lyme disease. Early treatment with antibiotics is usually effective. However, untreated, Lyme disease causes a persistent infection called late-stage Lyme disease that is difficult to treat.

Dr. Ken Singleton, an internist and active member of the International Lyme and Associated Diseases Society, is the author of *The Lyme Disease Solution*. He believes that many patients with chronic Lyme disease and related tick-borne diseases go on to develop abnormal endorphin levels that make them more susceptible to chronic inflammation and Lyme-related autoimmune disorders such as multiple sclerosis. Singleton, who has had complete recovery from his own case of Lyme disease, has found that increasing the body's ability to produce endorphins helps regulate the autoimmune dysfunction associated with Lyme disease.

Singleton explains that therapies that help improve and normalize endorphins are among the most promising therapies for immune regulation, inflammation management, and pain control. He has found that the best ways of increasing endorphins are with acupuncture, bodywork and massage, and low dose naltrexone, which has the best potential for restoring immune-system health.

In his practice, Singleton has found LDN to be beneficial for a number of his patients with Lyme disease. He attributes it to LDN's ability to correct immune system imbalance and restore the immune system's ability to regulate itself.[11]

Opioid Growth Factor and Bacterial Replication

Ian Zagon and Patricia McLaughlin have found that opioid growth factor (OGF), which is increased by low dose naltrexone, interacts with the opioid growth factor receptor (OGFr) to modulate the development of eukaryotes. Eukaryotes are organisms composed of cells containing a nucleus and an outer surface membrane. Specifically, Zagon and McLaughlin discovered that the interaction of OGF and its receptor controlled the growth of the bacteria *Staphylococcus aureus*, *Pseudomonas aeruginosa* and *Streptococcus marcescens*.

These results indicate that the complex formed by OGF and its receptor, known to be important in the regulation of mammalian development, also functions in the growth of simple unicellular organisms.[12]

At high doses opiate antagonists enhance and stimulate cell growth. Opiate antagonists can be used in high doses to accelerate the production of cells, including bacterial cells, to benefit fermentation processes used for the production of alcohol, vinegar and milk products, or to increase bacterial availability in sewage.[13]

Topical Creams for Pruritis

Preparations containing opiates and small doses of opiate antagonists have recently been developed to help reduce the incidence of opiate addiction in patients needing extended treatment with opiate-based pain medications. A side effect of these therapies is decreased incidence of pruritis (itching).

Because of this observation, researchers in Lausanne, Switzerland, have developed a topical cream containing 1 percent naltrexone, which is reported to be significantly more effective than placebo in relieving chronic itching in patients with allergic dermatitis. However, the naltrexone cream will not work in acute forms of itching, hives, or kidney-related itching.[14.]

Opiate Antagonist Therapy for Pruritus of Cholestasis

Increased opiodergic neurotransmission in the brain appears to contribute to the pruritis that accompanies certain chronic liver diseases. Studies show that because of this effect, administration of an oral opioid antagonist may precipitate a transient florid opioid-withdrawal type of reaction (characterized by anorexia, nausea, colicky abdominal pains, tremor, insomnia, perspiration, increased blood pressure, decreased pulse, cool skin, mood change, and hallucinations) in patients with pruritis related to liver disease. These effects are reported to resolve within two to three days when the low dose opiate antagonist is continued. In a collaborative study by researchers from the Academic Medical Center in Amsterdam, the Netherlands, Queen Elizabeth Hospital in Birmingham, England, and the College of Physicians and Surgeons, Columbia University in New York City, a low dose intravenous infusion of naloxone was administered prior to beginning oral naltrexone therapy to prevent this effect.

The dramatic reduction in pruritis associated with cholestasis attributed to opiate antagonists was first reported in 1979 and confirmed in clinical trials of the opiate antagonist nalmefene. However, the florid withdrawal reaction exhibited by some patients has complicated treatment. In addition to the use of naloxone infusion, the florid withdrawal reaction can also be reduced or eliminated by using clonidine along with low dose naltrexone or starting with a very small dose of naltrexone administered every three days.[15]

At Penn State, Ian Zagon continues his work with OGF, using RNA splicing (siRNA) to manipulate the OGF receptor, including the use of naltrexone with siRNAs. In March 2008, Zagon and his team conducted a study in which they used the cream Aldara to upregulate the OGF receptor, mediating the effects of OGF to treat skin lesions, including skin cancers. The study will be described in an upcoming edition of *Experimental Biology and Medicine*.

In another study expected to be released in the *Archives of Ophthalmology*, Zagon and his team have shown that topical opiate antagonist treatment of the diabetic eye of animals prevents neovascularization, a big problem that compromises vision.[16]

8

The Immune System and LDN in HIV/AIDS

The Developing Nations Project

Imagine a safe, inexpensive drug with the potential to prevent patients infected with human immunodeficiency virus (HIV) from progressing to acquired immune deficiency syndrome (AIDS). Better yet, imagine a collaboration of African and American physicians conducting a clinical trial of this drug in Mali, the former Sudanese Republic, in northwestern Africa. What has seemed for many a long-awaited dream has finally become real, and low dose naltrexone is the dream drug under investigation. In September 2007, as part of the Developing Nations Project, the First International LDN-HIV/AIDS Trial was approved by the Institutional Review Board in Bamako, the capital of Mali and commenced in December 2007.

This chapter provides an overview of the immune system and describes the potential contributions of LDN in HIV infection and AIDS. Here the reader will also learn about psychoneuroimmunology (PNI), an interdisciplinary approach that explains how anything that affects the immune system, the nervous system, or the endocrine system affects the other systems. In reviewing the effects of LDN, the contributions on each system toward restoring homeostasis must be considered.

LDN in HIV Infection

Studies show that low dose naltrexone has potent antiviral properties due to its ability to increase production of metenkephalins. In addition, LDN has been shown to increase levels of both endogenous opioid peptides and CD4 lymphocytes. Thus, LDN has effects that aid in restoring immune system health in patients infected with HIV. According to current theory, HIV targets specific immune system cells known as CD4+ T lymphocytes and causes certain immune system changes that compromise immune function. When these immune system changes become marked, patients develop AIDS and become susceptible to the development of opportunistic infections and tumors.

Studies dating from the mid–1980s, described in this chapter, show that by increasing the body's endorphin stores, LDN has the potential to strengthen and repair the immune system defects caused by HIV. As a consequence, patients with HIV/AIDS who are treated with LDN may experience improved immune system health and avoid disease progression. LDN exerts its primary effects by increasing production of the endorphin met-5-enkephalin. Metenkephalin has potent antiviral and antitumor properties due to its ability to increase production of interleukin-2, NK lymphocytes, CD4 lymphocytes, and CD 8 lymphocytes.

Protocol for the Mali Trial

The Mali LDN-HIV/AIDS Trial, expected to last nine months, involves patients on LDN alone, patients on LDN along with antiretroviral drugs, and patients on antiretroviral drugs alone. Study participants must be at least eighteen years old and have reduced CD4 lymphocyte counts in the 275–475 range at the onset of the study. The normal range for CD4+ lymphocytes is 550–1500.

Dr. Abdel Kader Traore and his colleagues at the University Hospital in Bamako are heading the Mali Trial. The hospital has an applied molecular biology laboratory headed by Dr. Ousmane Koita, a PharmD, Ph.D. graduate in clinical pharmacology from Tulane University. The National Institutes of Health and Tulane University provided most of the laboratory equipment. Two graduate students are helping with the study and writing their theses on the Mali Trial. In addition, five physician consultants, a virologist, an epidemiologist, a biostatistician, and several clinicians are involved with the project. Bernard Bihari is a consultant and helped design the study protocol. Irmat Pharmacy of Manhattan and Skip's Pharmacy in Florida have volunteered to supply LDN and placebo drugs for the trial at no cost.

Psychiatrist Jaquelyn McCandless, well known for her work with LDN in autism, and her husband, psychologist Jack Zimmerman, have been instrumental in bringing the LDN HIV/AIDS study to light. In addition, they volunteering their time and helping finance the study, which is being partially sponsored by the Ojai Foundation based in California. As a supplement to the medical aspect of the trial, Dr. Zimmerman is conducting a fifty-six week educational program designed to help reduce the incidence of HIV infection.

Why LDN Was Chosen

The choice of LDN in this study was based on the results of several scientific studies showing that low dose naltrexone can halt disease progression in HIV infection and improve immune system health. For instance, LDN has been shown to increase levels of both endogenous opioid peptides and CD4 lymphocytes,[1] the white blood cells targeted by the human immunodeficiency virus. Researchers speculate that its broad array of therapeutic effects enables LDN to reduce the incidence of opportunistic infections and malignancies.

In addition, LDN is inexpensive, easy to take, and free of adverse side effects. If manufactured in a developing country, the estimated cost of LDN would be no more than ten dollars per patient annually.

The Immune System and Disease

According to current theories regarding disease, most diseases stem from immune system defects caused by a number of different factors, such as oxidative stress or environmental agents (e.g. toxins, radiation), which lead to chronic inflammation or that affect cell growth and function. Consequently, many new therapies, both conventional and alternative, target the immune system. For instance, the drug etanercept (Enbrel), widely used in rheumatoid arthritis and other chronic inflammatory rheumatologic conditions, inhibits the pro-inflammatory immune system chemical known as tumor necrosis factor alpha (TNF-α). By inhibiting this compound, inflammation and pain are reduced.

In low doses, naltrexone induces increased production of metenkephalin, which has antiviral properties, influences cell proliferation, and helps restore homeostasis. One of the major benefits of LDN is its ability to increase the level of CD4+ T cells in patients with HIV infection.

The Immune System

The immune system is a complex network of organs and cells that work together to guard and protect us from infectious agents and toxins, and rid the body of damaged, infected, and malignant cells. A properly functioning immune system launches a meticulously orchestrated immune response whenever health is threatened by injury, malignancy, or infection. In autoimmune diseases, a weak, ineffective immune system mistakenly targets the body's own cells. Often, the symptoms of a particular disease or injury, such as fever and swelling, occur as a result of the immune response. For instance, when viruses enter the body, white blood cells rush to the infected area, releasing immune system chemicals that increase body temperature.

White Blood Cells and Opportunistic Infections

White blood cells, the major components of the immune system, are also known as leukocytes. As the protectors and defenders of the immune system, leukocytes constantly circulate throughout the body, scouting for foreign antigens. In the immune response, which is an example of homeostasis, white blood cells respond to any threat to health by increasing in number and rushing to the affected site.

Patients with leukemia, a condition of abnormally low white blood cells, or people who are deficient in certain types of white blood cells (for example, a deficiency of segmented neutrophils or neutropenia), have an increased risk of infection. The reduced CD4+ T lymphocyte cell counts in patients with HIV infection cause an increased risk of infection, particularly with opportunistic microorganisms. Opportunistic infections are caused by parasites and microorganisms, such as *Pneumocystis carinii*, that a healthy immune system is easily able to fight.

Leukocytes have two subtypes: granulocytes, which include neutrophils, eosinophils, and basophils, depending on the staining properties of the cellular granules; and mononuclear cells, cells with a distinct solitary nucleus, which include monocytes and lymphocytes.

Lymphocytes

Lymphocytes, the most important of the immune system cells, are a family of mononuclear leukocytes with several subtypes. The primary lymphocyte subtypes include:

- T-lymphocytes (T cells), which mature in the thymus gland where they're programmed to recognize specific antigens, which they're destined to find and attack; in response to an antigen attack, T-lymphocytes multiply in number.
- B-lymphocytes (B cells) are responsible for antibody production when they receive the appropriate command signal (antigen presentation) from T cells and macrophages.
- Natural killer (NK) cells are large, granular, cytotoxic lymphocytes that help eliminate infective or defective cells.

T-lymphocyte Subtypes

Several subtypes of T-lymphocytes exist. Each subtype

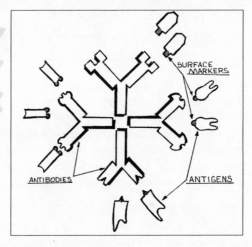

Antigens and antibodies (Marvin G. Miller)

has specific functions. Cell surface "cluster designation" (cd) markers on T cells are also associated with specific characteristics and functions. In good health, there is a normal balance of CD4 and CD8 T cells. CD4 T cells are decreased in HIV infection.

CD4 and CD8 T-lymphocytes

CD4 lymphocytes are known as helper inducer cells because of their ability to activate natural killer cells, macrophages, B-lymphocytes and other T-lymphocytes. CD4 T cells contain two subsets based on the type of cytokines they release: Th1 and Th2. Th1 lymphocytes are involved in cell-mediated immunity, for instance the destruction of infected cells by cytotoxic lymphocytes. Th2 lymphocytes are involved in antibody-related immunity, for instance the production of antibodies and autoantibodies.

CD8 lymphocytes are suppressor or cytotoxic cells that rid the body of cells that are infected or transformed by cancer. A reduced number of CD8 lymphocytes contributes to cancer development.

TH1 AND TH2 LYMPHOCYTES. Th1 and Th2 CD4 lymphocytes have different functions because they're each associated with different cytokines. Th1 T cells act to promote cell-mediated immune responses, whereas Th2 cells promote humoral (antibody-driven) immune responses. Th2 levels are increased in T-cell-mediated autoimmune diseases. The balance between Th1 and Th2 cells is critical for proper immune-system function. LDN causes an increase in metenkephalin, which increases levels of Th1 cells and reduces levels of Th2 cells.

Monocytes and Macrophages

Monocytes are large white blood cells classified as phagocytes, that engulf and destroy infected or damaged cells and infectious particles. Phagocytes can also present foreign antigens to lymphocytes, thereby initiating an immune response. Monocytes that migrate into the body's tissues become macrophages. Macrophages act as scavengers, ridding the body of aged cells and cellular debris.

Monocytes and macrophages are secretory cells in that they secrete immune chemicals known as monokines, as well as enzymes, complement proteins, and regulatory factors such as the cytokine interleukin-1 (IL-1). In addition, monocytes and macrophages carry receptors for cytokines, which are needed for activation.

Neutrophils

Neutrophils are the predominant type of white blood cell circulating through the blood. Neutrophils, which have segmented, multilobed nuclei, contain granules filled with potent chemicals that are released during the immune response, particularly during acute inflammatory reactions.

Dendrites

Dendrites, which are also called dendritic cells, are irregularly shaped white blood cells found in the spleen and other lymphoid organs. Dendrites have long, threadlike tentacles that enmesh lymphocytes and foreign antigens. Dendrites are also able to present antigens to lymphocytes and initiate an immune response.

Immune System Organs

The immune system is composed of a network of lymphoid organs. The bone marrow and thymus are called primary lymphoid tissue. All naïve immune system cells and the T-cell precursor cells arise in the bone marrow. Immature T-lymphocyte cells migrate to the thymus where they develop and mature.

Secondary lymphoid tissue includes the spleen, the lymph nodes, and associated lymphoid tissue present in the connective and organ tissues, including the skin, lungs, gastrointestinal tract and urogenital tract.

Lymph Nodes

Lymph nodes are aggregates of lymphoid tissue located strategically throughout the body. Lymph nodes serve as ambush points and filters for infectious and particulate matter.

Antigens and Antibodies

Viral and bacterial particles, as well as pollen and other proteins that the immune system cells react with, are called antigens. Antigens not normally present within the body are called foreign or non-self antigens. During the immune response, immune cells produce antibodies capable of destroying these foreign antigens. For instance, if someone is exposed to the human immunodeficiency virus (HIV), their immune system launches an immune response and it produces antibodies that help fight HIV infection.

In the immune response, antibody production is triggered when B-lymphocytes encounter or are presented with the antigens they're programmed to target. The B-lymphocyte then engulfs and digests the antigen and displays the antigen fragments bound or linked to its own distinctive marker molecules, known as major histocompatibility (MHC) markers, which in humans are called human leukocyte antigen (HLA) markers.

The combination of antigen fragment and marker molecule attracts the help of mature, matching T-lymphocytes. The T-lymphocytes then secrete cytokines that allow the B-lymphocytes to multiply and mature into antibody-producing plasma cells. Within the bloodstream, these antibodies lock onto matching antigens. These antigen-antibody complexes are quickly eliminated, either by the complement protein cascade or by the liver and the spleen. In certain autoimmune diseases, such as systemic lupus erythematosus, the antigen-antibody complexes can form lattice-like complexes that lodge in kidney tissue and interfere with normal kidney function.

Immune System Chemicals

The immune system produces a number of different chemicals, primarily cytokines and complement. These chemicals function as messengers, sending signals to other organs and cells that sustain or modify the immune response.

Cytokines

Cytokines are protein messengers such as the interferons and the endogenous opioids primarily produced by white blood cells, pituitary cells, and neurons. Cytokines can have variable actions depending on the type of cells they come into contact with. Cytokines can cause inflammation (pro-inflammatory cytokines); suppress inflammation (anti-inflammatory cytokines); encourage cell growth, activation and proliferation (growth factors); direct cellular traffic; destroy target cells; inhibit viruses and parasites; and incite macrophages. The primary effects of LDN involve its influence on cytokine levels.

The Nervous System

The nervous system includes the central nervous system (CNS), which includes the brain and spinal cord, and the peripheral nervous system, containing all neural elements outside of the CNS.

Nervous system cells are called neurons. The resident immune-system cells present in the nervous system are called glial cells, and include microglia and astrocytes. Nerves transport messages and signals from the body's organs, including its lymphoid organs and cells, to and from the brain and other organs. The neural network exerts its influence on the immune system via its effects on the endocrine system.

Effects of LDN on Neurodegeneration

The death and destruction of neurons occurs in a process known as neurodegeneration. Some diseases, including Parkinson's disease and multiple sclerosis, are considered the result of neurodegeneration. By blocking the activity of excitotoxins (chemicals that destroy neurons), LDN prevents destruction of oligodendroglia, which are precursor glial cells, and oligodendrocytes, the neural cells that produce myelin found in the protective nerve sheath covering nerve fibers. This helps prevent myelin destruction in patients with multiple sclerosis and helps prevent central nervous system damage in patients with HIV infection. For more information on the effects of LDN on neurodegeneration, see chapter 4.

Neuropeptides

Neurons use chemical signals in the form of neurotransmitters, peptides, and gases such as nitric oxide to communicate information. Neurons produce specific peptides and neurotransmitters, depending on their location within the central nervous system. Peptides are long strings of up to fifty amino acids that act as informational proteins or messengers. Peptides that function within neural tissue are called neuropeptides.

Neuropeptides, which include somatostatin, oxycytocin, vasopressin, growth hormone, prolactin, and insulin are synthesized within neurons under direction of DNA within the cell nucleus. Neuropeptides act at receptors of the G-protein family expressed by selective populations of neurons. As a result, peptides communicate signals between one population of neurons and another. When cells are stimulated, they release neurotransmitter molecules from large cell vesicles in the cell body through the axon onto the terminal of the cell. Neuropeptides have diverse effects, including gene expression, local blood flow, and glial cell morphology. Neuropeptides tend to have prolonged actions, and some have striking effects on behavior.

Neurotransmitters

Neurotransmitters, which include neuropeptides and classical transmitters, were originally described as chemicals relaying signals between neurons and other cells. Now it's known that leukocytes can also release neurotransmitters capable of affecting surrounding cells.[2] White blood cells are known to have receptors for the classical neurotransmitters dopamine, serotonin, norepinephrine, and beta-adrenergic compounds.

The classical transmitters, which also include epinephrine, histamine, gamma-aminobutyric acid, glycine, and glutamate, are small, low-molecular-weight compounds. Neurotransmitter molecules attach to a receptor on the membrane of the nerve fiber causing a change in the fiber's shape, which opens a pore that allows sodium or calcium ions to pour inside the axon's synaptic cleft, which triggers the cell to fire and transmit a signal down its axon fiber in a process of exocytosis. Neurotransmitters affect the excitability of other neurons by depolarizing or hyperpolarizing them.

The Opiate Receptor

Opiate binding sites (opiate receptors) in the central nervous system were first proposed in 1954 and demonstrated in mammalian brain tissue by Candace Pert in 1973. The opiate receptors are now known to be widely distributed in the central nervous system and in peripheral sensory and autonomic nerves as well as immune system cells. When the opiate receptor is activated by either endogenous or exogenous opiate compounds, a number of physiological functions and behaviors are initiated.

Signal Transducers

As G protein cell receptors, opiate receptors are signal transducers anchored to the cell surface plasma membrane. Here they connect receptors to effectors and thus to intracellular signaling pathways. When an agonist such as morphine binds to an opiate receptor, the G protein is split into two subunits, G alpha and G beta gamma. Both of these subunits activate intracellular, second-messenger systems regulating cellular components such as metabolic enzymes, ion channels and the transcriptional machinery. Changes in potassium and calcium channels, in turn, lead to reduced excitability and inhibition of neurotransmitter release. Functionally, the endogenous opioids act as co-transmitters modulating the effect of fast-acting neurotransmitters.

A reduction in receptor binding may occur as a result of receptor downregulation, delay in receptor recycling, or a pathological change related to neuronal damage or neurodegeneration.

PET Scan Analysis

Opiate receptor binding can be studied with positron emission tomography (PET) scanning. Since 1985 such studies have been applied for neurochemical mapping and for assessments of pain, emotion, drug addiction, movement disorders, neurodegeneration, and epilepsy. Such studies have shown changes in receptor binding for men and women as well as changes in receptor binding in different areas of the brain related to age, circulating estrogen levels, and health status. For instance, dysfunctions of opiate receptor binding in the basal ganglia are seen in movement disorders. However, no changes are seen in patients with Parkinson's disease without dyskinesia when compared to healthy adults.

The Endocrine System

The endocrine system is a network of glands that produce hormones. The endocrine system influences the immune system by producing cytokines and hormones with direct effects on immune function. Conversely, leukocytes produce hormones and cytokines that affect endocrine function.

Endocrine glands include the hypothalamus, pituitary, thyroid, adrenal, parathyroids, thymus, ovaries, testes, pineal, and the islet of Langerhans in the pancreas.

Hormones

Hormones are messenger proteins. Glands release hormones directly into the bloodstream and direct them to distant organs or cells where they exert specific actions by binding with cell receptors. Hormones may also have a paracrine effect, in which they influence neighboring cells, and an autocrine effect in which they exert action on the cells that produced them.

Neurons and also non-neural cells are known to secrete various releasing hormones.[3] Releasing hormones can have various functions. For instance, corticotropin-releasing hormone causes the release of corticotropin from the anterior pituitary gland, and it can function as a modulator of neural activity in other brain regions or act as a cytokine in lymphoid tissue.

The Pituitary Gland

The pituitary gland is considered the master gland because its hormones influence other endocrine glands. The pituitary's anterior and posterior regions produce different hormones. The anterior pituitary is of particular importance to immune function due to its release of the hormones prolactin, luteinizing hormone, growth hormone, and the cleavage products that result from the breakdown of proopiomelanocortin (POMC) protein, including adenocorticotropic hormone (ACTH), which is co-released with the endogenous opioid peptide beta-endorphin.

The Adrenal Gland

Stimulated by ACTH, the adrenal gland releases glucocorticoid hormones that have the effect of suppressing the immune system. The adrenal gland also releases epinephrine and norepinephrine into the circulation system.

Growth Hormone and Prolactin

While all of the endocrine hormones influence immune function, growth hormone produces some of the most powerful effects, including an ability to increase the size of lymphoid tissues, particularly thymus tissue. All of the immune system cells have receptors for growth hormone with the highest number of receptors seen on B-lymphocytes and macrophages, and the lowest number of receptors seen on T cells.

Prolactin, which has structural similarities to growth hormone, also facilitates immune responsiveness. Prolactin plays an important role in the activation of macrophages, and an inhibition of prolactin is associated with a poor immune response and an increased lethality in infections.

The Pituitary-Adrenal Axis: POMC Peptides and Glucocorticoids

The release of corticotropin-releasing hormone from the hypothalamus causes the release of the POMC peptides, adrenocorticotropic hormone (ACTH) and the co-release of beta-endorphins. ACTH modulates immune function by decreasing antibody production, interfering with macrophage-mediated tumor destruction, and by suppressing cytokine production. Beta endorphins have immunomodulating properties, increasing antibody production when it is low and suppressing it when it is high.

Psychoneuroimmunology

Psychoneuroimmunology (PNI), a term also referred to as neuroimmunology, was first described in 1975 by Dr. Robert Ader at the University of Rochester. PNI refers to the study of the interactions that occur among the nervous system, the immune system, and the endocrine system. For instance, the bone marrow, thymus, spleen, and the lymphoid organs are innervated or activated by the autonomic nervous system, and lymphocytes of the immune system produce neuropeptides that affect the neural system. In addition, hormones and neurotransmitters produced by the endocrine system can affect behavior as well as immune function.[4]

Communication between the brain, the endocrine glands, and the immune system occurs through signaling pathways, including the hypothalamic-pituitary-adrenal axis (HPA) and the sympathetic nervous system. The HPA is particularly important in the stress response. Here, inflammatory cytokines produced during the immune response stimulate adrenocorticotropic hormone and cortisol release.

Endogenous Opioid Peptides

Many effects of LDN are attributed to its ability to transiently increase production of endogenous opioid peptides. Endogenous opioid peptides are naturally occurring proteins produced by neurons, immune system cells and pituitary cells with analgesic and euphoric properties similar to those of opiate drugs. However, their effects on the immune system are markedly different. While opiates are known to suppress immune function, the endogenous opioid peptides Beta (β)-endorphin and [Met-5]-enkephalin (metenkephalin, MET) increase NK cell activity and lymphocyte proliferation. Endogenous opioids are also able to function as growth factors and they have variable actions depending on their dose and their duration.

Three distinct endogenous opioid peptide families have been identified: enkephalins, endorphins, and dynorphins. Each of these subtypes is derived from a distinct precursor protein and has a characteristic anatomical distribution in the body.

LDN or Metenkephalin

Research on opioid peptides is fundamental to understanding the interrelationship between the neuroendocrine and immune systems. In 1979 Wyban and colleagues made one of the first observations of this interaction.[5] They showed that morphine, a drug once considered to only affect neuronal function, could also affect the responses of human B-lymphocytes, thereby affecting immune function. This led to other experiments and the discovery that several neuropeptide receptors exist on immune sys-

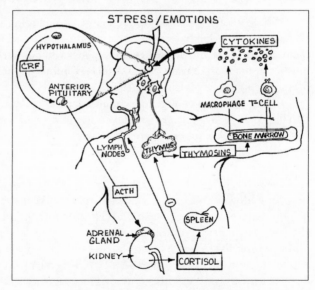

Stress emotions (Marvin G. Miller)

tem cells. This, in turn, showed that many, and possibly all, neuropeptides can affect the immune response.

Variable Actions of Opiate Compounds on Opiate Receptors

The cloning and subsequent identification of the genetic coding regions for opioid receptors have contributed significantly to our understanding of opiate receptors. The ability of a drug to bind with its receptor and cause a unique action is no longer considered specific. As an example, opiate receptor identification studies have shown that the ability of two opioid receptors, for instance mu and delta, to interact and form unique structures in a process of dimerization may alter the pharmacological properties of drugs that bind to them.[6]

These studies also show that drugs that are relatively selective at standard doses will interact with additional receptor subtypes when given at sufficiently high doses, which leads to possible changes in the drugs' pharmacological actions.

Persistent or Intermittent Blockade of the Opiate Receptor

Opiate antagonists and opiate agonists, including endogenous endorphins, react with opiate receptors that reside on the nucleus or on the surface membrane of cells, much like keyholes in locks that opiate agonists and antagonists can open or activate. Various factors influence the specific effects that occur when the receptor is activated.

In his studies on endogenous opiates, Ian Zagon has found that intermittent blockade of the receptor with low doses of opiate antagonist drugs upregulates production of endogenous opiates. Endogenous opiates such as metenkephalin and beta endorphin regulate homeostasis, immunity, wound healing and cellular removal.

However, persistent blockade of the opiate receptor and persistent elevations of endogenous opiates, particularly metenkephalin, contribute to cell growth. Several studies of hepatic encephalopathy, hepatitis B, and hepatitis C show that higher levels of metenkephalin in the liver are associated with more severe liver damage, cirrhosis, and the development of hepatic encephalopathy. Persistent blockade stimulates cell growth, including cancerous cell growth, whereas intermittent blockade with LDN has an inhibitory effect and restores homeostasis.

Along this line, low dose naltrexone has been shown to offer benefits in hepatic encephalopathy, viral hepatitis, and cholestasis-induced liver injury.[7]

Beta-endorphin

Beta (β)-endorphin is the major opioid peptide derived from the precursor protein preproopiomelanocortin (POMC). Adrenocorticotropic hormone (ACTH) and melanocyte-stimulating hormone (MSH) are also derived from POMC. Preclinical evidence shows that endogenous opioid peptides alter the development, differentiation, and function of immune cells by reacting with opioid receptors on immune system cells.

The Emergence of Acquired Immune Deficiency Syndrome (AIDS)

In 1980 a marked increase in reported cases of the rare disorders Kaposi's sarcoma and *Pneumocystis carinii* pneumonia alerted scientists to the possibility of a new disease. The first case of this disorder, which eventually became known as acquired immune deficiency syndrome (AIDS), was reported in 1981. By 1984, researchers had identified the human immunodeficiency virus (HIV) as the cause of AIDS. Early on, researchers also discovered that T-lymphocytes, specifically T-lymphocytes with surface markers for CD4, were targeted and destroyed by either the HIV virus or as a consequence of the immune system's response to HIV.

LDN and the Effects of Increased Endorphins

LDN has been shown to increase levels of beta-endorphin and enkephalins. Beta-endorphin and other opioid peptides appear to act as immunomodulators capable of suppressing or enhancing immune responsiveness. For instance, in response to antigenic stimulation, beta-endorphin increased low antibody production and decreased high antibody production.[9]

Bernard Bihari's Studies

In the early 1980s Bernard Bihari noted a high rate of AIDS cases among the heroin addicts he was treating. Aware of Ian Zagon's studies on LDN and the considerable laboratory studies showing the central role of endorphins in regulating immune function, Bihari began conducting clinical trials of LDN in AIDS patients in 1985–86 (see below).

In his practice, Bihari and his staff have treated a number of AIDS patients with LDN alone who have experienced only a minimal decline in immune function. In an interview with Kamau Kikayi in 2003, Bihari emphasized that LDN can stop disease progression in HIV infection as long as patients do not become reinfected.[10]

The First Clinical Trials
for LDN in AIDS

Bernard Bihari and his colleagues conducted a placebo-controlled trial of LDN in 1985–86 in thirty-eight patients with AIDS. Bihari measured the endorphin levels of these subjects and found that they were less than 25 percent of the normal range. Subjects were given a 1.75-mg dose of LDN taken orally at bedtime. LDN is known to block the mu opiate receptor for approximately three hours. Nearly all of the body's endorphins are produced by the adrenal and pituitary glands between 2:00–4:00 A.M. Following the blockage, the body's production of beta-endorphin and metenkephalin increased for an average period of twenty to twenty-four hours.

In this trial Bihari also found that a dose of 3 mg increased endorphin levels in the study participants by 200–300 percent. He also found that doses lower than 1.5 mg had no effect on endorphin production. Doses higher than 4.5 mg did not raise endorphin levels higher than lower doses. However, higher doses blocked the mu opiate receptor for a longer period, which reduced the benefit of the increased endorphin levels.

The trial, which initially lasted twelve weeks, showed a significant reduction in the incidence of opportunistic infections. Of the sixteen patients receiving a placebo, five opportunistic infections occurred, whereas in the LDN

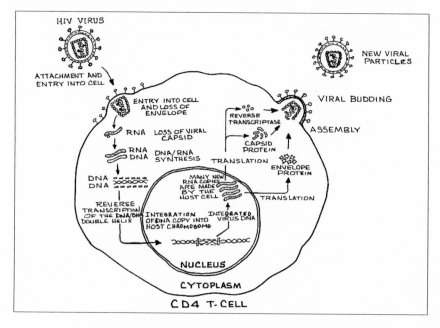

HIV infection (Marvin G. Miller)

treated group, no opportunistic infections occurred. Following these initial trial results, the placebo group was given LDN and both groups were observed for another three months (the trial lasted from July 30, 1985, through January 15, 1986).

Both groups continued to be monitored by Bihari after the trial's end. By December 17, 1986, 87 percent of patients in the original placebo group had died, while 17 percent of patients in the LDN group had died. None of the patients were given prophylactic antibiotics to prevent opportunistic infections during the course of the trial although this is considered standard therapy today. It should be noted that many of the participants in the trial had full-blown AIDS at the onset of the study.[11]

Benefits of LDN in AIDS

Laboratory tests performed in conjunction with the 1986 trial indicated that lymphocyte mitogen responses declined in patients receiving a placebo drug, but they did not decline in patients receiving LDN. In addition, pathologically elevated levels of the cytokine acid-labile alpha interferon, which were present in all patients, declined significantly in the patients taking LDN but not in the placebo group. When the entire group was switched to LDN, 61 percent of patients experienced a decline in alpha interferon levels. Bihari concluded that a decrease in alpha interferon appears to be the crucial factor in patient stabilization. Evidence suggests that the high alpha interferon levels seen in patients with AIDS contribute to immunosuppression.[12]

LDN and Antiretroviral Drugs

Since the introduction of highly active antiretroviral therapy (HAART) for AIDS, Bihari has used LDN in combination with HAART (indinavir and nevirapine) for nearly all of his AIDS patients with lower CD4 T cell counts. Bihari's patients using LDN with HAART have experienced an 87 percent increase in CD4 T cell counts, which is significantly higher than that seen with HAART alone.[13] Seventy-five percent of these patients have also exhibited the slow secondary rise in CD4 T cell counts that occurs at about eighteen months after starting HAART treatment.

In addition to Bihari's ongoing studies, researchers at the Institute for Brain and Immune Disorders, Minneapolis Medical Research Foundation, Hennepin County Medical Center and the University of Minnesota Medical School in Minneapolis conducted an in vitro study with CD4+ T-lymphocytes obtained from HIV-1 seronegative individuals.

The cells were infected with HIV-1 and incubated for four days with AZT, indinavir, naltrexone or combinations of the drugs. Levels of HIV-1 p24 antigen were measured. The researchers found that naltrexone alone had no effect on HIV-1 expression by the cells. However, when naltrexone was added to cell cultures containing AZT or indinavir, the antiviral activity of these drugs increased. The researchers concluded that naltrexone appeared to target the infected cells rather than the virus itself, making it a novel effective treatment approach.[14]

Effects on Lipodystrophy

Lipodystrophy is a condition of peripheral wasting with a loss of subcutaneous fat in the face, arms, legs and buttocks, and an accumulation of abdominal visceral fat that may be accompanied by the presence of a dorsal fat pad (buffalo hump), elevated lipid levels, and insulin resistance, that can occur as a side effect of antiretroviral drugs. Patients with HIV using low dose naltrexone in conjunction with antiretroviral drugs are reported to remain free of lipodystrophy, suggesting that LDN offers protection from this disorder.[15]

Follow-up of Bihari's Patients

As of November 2007, Bihari had treated 350 AIDS patients using LDN either alone or with antiretroviral drugs. More than 85 percent of the patients currently undergoing LDN treatment for HIV showed no detectable levels of HIV and no significant side effects.[16]

Study of Metenkephalin in ARC

At the April 2007 National Cancer Institute Conference on low dose opioid blockers, Nicholas Plotnikoff presented an overview of metenkephalin's actions in HIV/AIDS. Plotnikoff described the effects of LDN, which include diminishment of viral breakthrough and restoration of CD4 T cell levels.

Plotnikoff also described the results of a twelve-week placebo-controlled trial that he and Bihari conducted using two doses of intravenous methionine enkephalin (metenkephalin,) in forty-six patients diagnosed with AIDS-related complex (ARC). In this trial, patients using higher doses (125 u/kg/week) produced significant increases in interleukin-2 (IL-2) receptors, CD56 natural killer cells, pokeweed mitogen-induced blastogenesis, lymphocyte percentage, and CD3 cell numbers compared to baseline levels, and a significant increase in CD4 and CD8 T-lymphocytes compared to patients in the placebo group. In addition, no signs of toxicity were observed and patients showed a significant reduction in total lymph node size. The results suggest that metenkephalin is safe and may have a beneficial immune modulating effect in patients with ARC.[17]

Follow-up of Mali Trial

McCandless and Zimmerman report that they have begun training with volunteers in preparation for counseling sessions on gender and sexual relationship issues that are being held in conjunction with their naltrexone trial. The need for counseling is apparent from a recent UN study showing that 85 percent of Mali women indicated that they did not feel empowered to make decisions that affect their intimate life. McCandless has set up a Web site, *www.ldnafricaaids.org,* where she is posting a blog and information related to the trial.

9

The LDN Experience:
A Patient's Guide to LDN

SammyJo Wilkinson is a patient activist who compiled the practical information on LDN presented in this chapter. In addition, she shares her story of being diagnosed with crippling multiple sclerosis (MS), and she describes the amazing improvements she's experienced since starting LDN. Because it is pertinent to her story, Wilkinson also describes the contributions of the Internet and patient activists in spreading the word about LDN, and their success in initiating the first clinical human trial of LDN therapy for MS in the United States. The Internet is where Wilkinson first learned about LDN.

In researching the practical information about LDN presented in this chapter, Wilkinson consulted with several experts in the field to confirm the accuracy of the information presented here. This information is intended for anyone considering using low dose naltrexone (LDN) or who would like more information before approaching their physician about it. Topics covered include:

- considerations in choosing LDN
- information to share with your doctor
- need for patient empowerment
- the status of LDN research
- sources of LDN and doses
- LDN's effects, safety, side effects and contraindications.

The suggestions in this chapter are not meant to replace the advice of one's physician. Rather, they are intended to serve as an educational guide to the use of LDN, and to empower patients and assist them in making informed health-care choices. Armed with information, patients will be better equipped to discuss LDN with their health-care providers.

It's important for patients to work closely with their physicians to ensure that their condition is monitored and that LDN's safety and efficacy for their condition is carefully evaluated. While LDN is reported to be safe, it's still considered an experimental, off-label treatment for conditions other than drug and alcohol abuse, with human clinical trials for other conditions still in their beginning stages.

Having your doctor support and cooperate with your decision to try LDN not only ensures an ongoing positive relationship with him or her, it also makes you a member of the pioneering LDN patient community for having educated one more patient provider. The next patient who requests LDN from your doctor is likely to receive an informed response. And after your doctor observes your progress on LDN, he or she may even become proactive in suggesting it to other patients.

SammyJo Wilkinson's LDN Story

In 1995, at age thirty, I began to experience numbness and burning in my hands and feet. After enduring a series of diagnostic tests, I was diagnosed with multiple sclerosis. My disorder was classified as relapsing remitting MS, which means that my episodes of symptoms alternated with periods of remission. During the first few years of my illness, I had violent attacks of MS every few months. During these attacks I was either unable to walk or experienced vision loss. After each attack, I would recover enough to go back to work, but there would always be some lingering problem that added to the growing list of physical impairment and symptoms.

From the long list of MS symptoms found in the medical literature, it seems I've had nearly all of them. After my diagnosis, numbness and tingling increased until my fingers became so numb I frequently dropped things. Optic neuritis appeared early on, blurring my vision and causing black spots referred to as "floaters." My balance was also affected early, leading to an unsteady gait, progressing over the years and leading to dangerous falls. Urinary urgency and constipation, common symptoms in MS, became increasingly difficult to manage. Stiffness in the morning made it hard for me to leave my bed, and spastic leg jerking at night made it hard to sleep. Heat intolerance prevented me from stepping outside if the temperature exceeded 76 degrees. Cognitive function was impaired and characterized by memory problems, confusion and "brain fog."

Cognitive problems even affected my speech. I'd forget the names of common objects, or I would use the wrong words, saying "hammer" when referring to a wrench that I was holding. Sensory pins-and-needles feelings, as well as numbness, appeared and disappeared across various parts of my body. When numbness spread to my pelvic region, I was left with sexual dysfunction, due to lack of feeling.

Fatigue was a major impediment from the onset, greatly limiting my activities. Fatigue worsened with the slightest effort. After a few steps my legs invariably felt like cement and I could barely lift my feet. Fatigue, combined with gait disturbances and needing to run to the bathroom every fifteen minutes, eventually kept me confined to my home at the height of my MS disability.

By 2002, I had advanced to secondary progressive MS, and was forced to quit my job as a technology CEO. After 15 years in Seattle, my husband and I decided to move back to Texas where I had family.

Failed MS Treatments

Over the course of my disease I was prescribed drugs from every major category of MS medication available. Initially, intravenous (IV) infusions of corticosteroids (such as methylprednisolone) were the only therapy available to calm my attacks. I anxiously awaited FDA approval of the first drugs specifically for MS. Supporting studies reported around a 30 percent reduction in attacks and slower disease progression.

In 1998 I started injecting Copaxone daily. Initially, the injections were tolerable, with the needles only half an inch long, injected subcutaneously into thighs, stomach and arms. However, these constant injections eventually resulted in large and painful purple bruises. Over the years, my legs became lumpy with scars, and the skin on my stomach started to turn gray; my doctor called it necrosis and told me the skin was dying. The scars became hard with fibrosis, and made it difficult to get the needle in on the first try.

Despite my adherence to this approved protocol, my condition continued to worsen. By 2002, my neurologist told me that, as I suspected, I wasn't responding to the Copaxone. However, I was alarmed to learn that my diagnosis had changed and that I now was classified as having secondary progressive MS. My doctor said that my last resort was the chemotherapeutic drug mitoxantrone (Novantrone), recently approved to treat worsening MS, taken intravenously. The listed side effects were ominous, including a small chance of fatality related to the development of leukemia or heart failure. Apprehensive and frightened by the thought of my disease continuing to progress, I felt I had no choice.

I received the first three IV doses of mitoxantrone monthly. These were to be followed by an additional dose every three months until the two-year course was complete. By the end of the first year I had received seven doses, and it was taking longer and longer to recover from the treatments. Worst of all, it was clear that my MS was still worsening. My right leg became partially paralyzed, and so atrophied that it was visibly thinner than the left leg. I started using a cane to walk because I was so weak and unsteady.

I was also required to have my heart monitored while on this drug, via a MUGA scan, which entails extracting blood, irradiating it and reinjecting it to monitor the heart's function while it pumps. When the MUGA scan came back reporting my heart function was down ten points, I had to stop taking the drug before completing the two-year therapy course. I was left with a damaged heart, the veins in my arms scarred and riddled with tracks from the harshness of the drug, and I was now disabled by MS. The only perk was that I didn't have to walk so far after parking the car, thanks to the blue tag hanging on my rear view mirror.

MOVING ON. It was at this point that we gave up our life in Seattle, the beautiful new home my husband Doug had built, and the computer company I had founded. We moved back to my home state of Texas. My relatives were shocked by how much I'd declined and feared that I wasn't going to make it much longer. I was having the same thought. Our lives were in chaos from my health and the cross-country move, but my husband Doug kept his bright outlook to cheer me on because he could tell I was close to giving up.

I found a new neurologist at the University of Texas MS Center, but, having been through so much, I had little left in the way of treatment options. Having failed two of the approved therapies for MS, my doctor had no choice but to suggest the third category of approved MS drugs, the interferons.

Consequently, I began injecting Avonex weekly. This was complemented with monthly IV corticosteroid injections. Needles and infusions had been a way of life for years, but now I was performing deep intramuscular self-injections weekly, using a needle twice as large as the needle I'd used for injecting Copaxone.

Since my initial diagnosis of MS I'd also been prescribed several oral medications to relieve symptoms, including oxybutynin for bladder control, and baclofen for muscle stiffness. In Texas my new neurologist added gabapentin (Neurontin) to reduce muscle spasms, and Wellbutrin to help alleviate my depression.

A home health nurse visited me each month to administer the IV corticosteroid infusions. This was quite a task, with my scarred veins. The nurse had to probe my arms to find an accommodating vein. Unfortunately, my veins started to scar from the monthly injections. If the nurse found accommodating veins, they would invariably rupture before the infusion was complete and leave large bruises and hematomas.

Shortly into the new program, my assigned nurse left in despair after failing to find a usable vein despite nine separate attempts. The home health service switched me to their neonatal nurse who specialized in IVs for the delicate veins of premature babies. Although my veins benefited from the switch, unfortunately none of these therapies provided any relief of my symptoms. The complications that accompanied these injections added to the stress of dealing with MS. In addition, I began getting flu-like symptoms from the Avonex, which compounded the MS fatigue.

I'd also become intolerant of the corticosteroids. They no longer reduced symptoms or increased my energy the way they did when I first started using them. The nurse had already recommended to my doctor that I get a surgically implanted venous port if I were to continue infusions because my veins were no longer cooperating. Still, my MS condition deteriorated to the point that I was falling so often I had no recourse but to follow my doctor's earlier recommendation. I ordered a $10,000 motorized wheelchair.

A Glimmer of Hope

In desperation, I took one more troll on the Internet, and typed in "MS cure" even though I knew I had an incurable disease. My online MS searches had previously always disappointed me. Usually, I found scams designed to take advantage of desperate people, or hints of medical research promising new treatments maybe forthcoming in the next decade but too late for me.

This time I found something different. I read several posts in discussion forums about a drug called low dose naltrexone or LDN. MS patients taking LDN claimed to be improving. Nobody said it was a cure. But even people with the worst kind of MS, primary progressive, reported that they were getting relief from symptoms, and they claimed that their disease progression appeared to have stopped.

Naturally, after all I'd endured, I was skeptical that something as simple as a pill taken orally once daily could help me. But as I continued surfing the Net, I learned that LDN's main advocate was a Harvard-trained neurologist practicing in New York City, Bernard Bihari. Could this physician and thousands of patients reporting their anecdotal success stories online be wrong?

Cautious by nature, I continued to research LDN online for two more months before I decided to give it a try. Of course, my neurologist at the University of Texas said he'd never heard of LDN. He refused to prescribe it. By then I was on a mission. After inquiring online, another patient in the forum gave me the name of a doctor who would prescribe LDN.

My LDN Experience

I took my first 4.5-mg capsule February 4, 2003, at bedtime. I woke in amazement eight hours later. This was the first time in years I had slept straight through the night without leg spasms or having to get up multiple times to go to the bathroom. My husband was so used to my fitful sleeping that he didn't sleep a wink that night, worried that I was so still.

I watched my symptoms and strength carefully, afraid to get my hopes up. By four weeks of taking LDN, I knew I had more energy. I could stand at the stove long enough to cook breakfast. At six weeks there was no question. I no longer needed my cane.

Soon after, I went for an appointment with the neurologist, and Doug mischievously set him up. "Doc, what are the chances Sam will be able to avoid the wheelchair and walk without her cane?" My doctor answered, "Don't count on it." Then I laid down the cane and walked down the hall, while my doctor watched in disbelief.

Next Doug said "What are the chances she could jump again?" The doctor was getting suspicious at this point, but he said "I've never seen a secondary progressive patient achieve significant recovery." At which point I jumped six inches off the ground, and confessed I'd started LDN. A broad smile broke over my doctor's face as he whipped out his prescription pad and said, "How do you spell it, I'll write the script!" I went home and canceled that wheelchair.

As winter turned into spring, my symptoms continued to improve rapidly. In April I injected my last dose of Avonex, and I was able to start weaning off all of the oral medications. One of the earliest changes I noticed was that my vision problems and cognitive deficits started clearing. The tingling and numbness also resolved. One morning I woke up and had feeling in my fingers again. Even the sexual dysfunction of several years resolved as my sensory functions returned. Needless to say, my mood was soaring after the years of no hope. Initially I worried that I might be experiencing a massive placebo effect, but my recovery never faltered.

The previous Texas summer I had spent huddled in front of the air conditioner. Heat makes MS patients fall apart because of their unshielded nerves. Now I found my heat tolerance increased. By May I was pulling weeds in the garden in the 85-degree heat. The fatigue lifted, and I was able to participate in more physical therapy regimens.

Before I started LDN my therapist had to keep fans on me and have ice packs handy, as even mild exertion led to immobility from overheating. Now I was able to do more and more every week at the clinic, amazing all the therapists as I regained functions that had been lost for years. My balance and walking distance improved, as did urinary urgency. We celebrated by taking our first long excursion away from home, a several-hundred-mile trip to Big Bend National Park down on the Rio Grande. By September, my husband said, "Sam, I think you're going to make it. Let's go back home to Seattle!"

No More Needles

My MS history, treatments and improvements after starting LDN are summarized in the following chart, using the standard Expanded Disability Status Scale (EDSS), since that is what the neurologists used to track my decline. The scale begins at 0, indicating no disability, and ends at 10, death. The need for a power chair placed me at 7, just before I found LDN.

I no longer live in a constant state of MS misery, and needles are no longer a part of my life. Most of my symptoms have been gone since 2004, including the overall sick feeling I had that I've been told was caused by a chronically inflamed immune system. After the first year of rapid improvements, my recovery leveled off. There is still a "hitch in my giddyup," as they say in Texas, but despite an imperfect gait, I get around with no assistance. I still attend physical therapy sessions regularly and I practice movement therapy to maintain

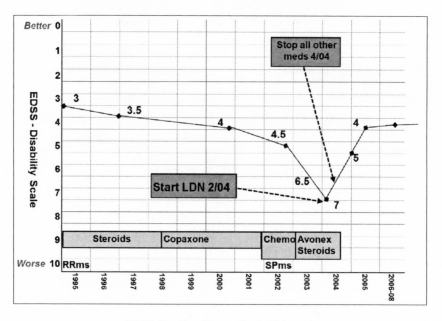

EDSS timeline (Samantha Wilkinson)

what I've regained, and continue improving. The adage "if you don't use it, you lose it!" holds true when it comes to maintaining what I've regained with LDN.

The disabilities I am left with are harder to overcome than the erratic symptoms caused by inflammation. Scarring accumulated in my central nervous system over the years, before the damaging disease activity was halted. This damage is very slow to repair itself, but researchers have established the possibility that it can. I remain optimistic, especially with the MS-related inflammation under control with LDN.

LDN Research

Online, by early 2003, I was becoming one of those with amazing anecdotes, excitedly comparing recovery notes with others who had been using LDN longer, and advising and supporting others new to LDN. But I was frustrated by the lack of research that had been done on LDN in MS. Dr. Glen Gordon, a physician friend, encouraged me to put my computer skills to use and launch my own research effort by tracking patient responses. I demurred, thinking, "Who would pay attention to a non-scientist?" But Gordon said that "when the n (the number of subjects in a study) becomes large enough, it's no longer anecdotal, it's epidemiologic data."

Online, I met another doctor, Yash Agrawal, in the LDN forum. He was working on a theory about how LDN actually works in MS, and told me what it would take for research to become reality: a hypothesis, which he eventually published in *Medical Hypotheses* (see chapter 4); animal studies to show LDN's safety and efficacy in MS; and clinical trials in humans.

We started contacting university researchers, telling them about the rich store of anecdotal evidence and the hypothesis Dr. Agrawal had designed. We couldn't interest any scientists unfamiliar with LDN. However, as we continued searching online we learned that Ian Zagon at Pennsylvania State University had been researching LDN for more than twenty years. In fact, Dr. Zagon had published hundreds of studies on opiate antagonists and their benefits to human health.

We also learned that Dr. Zagon's studies had first inspired Dr. Bihari to use LDN in his medical practice. Then we found that Jau-Shyong Hong at the National Institute of Environmental Health Sciences (NIEHS) at the National Institutes of Health (NIH) was conducting research on naltrexone in neurodegenerative diseases. Now I was convinced that the time for studies of LDN for MS had come. But how could I get the attention of researchers?

My Online Survey

In an effort to convert all the wonderful anecdotal reports I was hearing into meaningful statistics, I launched my own humble research project, an online LDN survey for MS, which can be found in Appendix B, categorized by the four types of MS. I also created a Web site, LDNers.org, to host the survey and offer information specific to LDN's use in MS. Yash Agrawal and I designed the questionnaire, and he helped me analyze the 267 responses collected from patients worldwide. The results showed:

- a very low relapse rate of 0.23; one patient experienced relapse in five years
- 70 percent of patients reported symptom improvement
- 45 percent of patients thought that their disease progression had stopped
- 76 percent of patients reported that LDN was working and that they planned to continue using it.

In order to understand how significant the low relapse rate reported by the LDN survey is, the following chart compares relapse rates reported for the three primary FDA-approved MS treatments. The benchmark is the untreated patient, who typically experiences one relapse per year.

The data source for the relapse rate shown for no treatment, Avonex, Betaseron and Copaxone is a study titled "A Prospective, Open-label Treatment Trial to Compare the Effect of IFNbeta-1a (Avonex), IFNbeta-1b (Betaseron), and Glatiramer Acetate (Copaxone) on the Relapse Rate in Relapsing-remitting Multiple Sclerosis: Results after 18 Months of Therapy," published in the journal *Multiple Sclerosis*, December 2001.

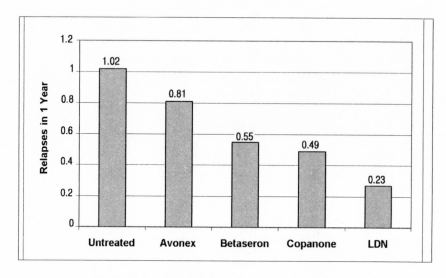

Relapse rates in multiple sclerosis (SammyJo Wilkinson)

The data from this ABC (Avonex, Betaseron, Copaxone) drug study show that my survey participants reported a relapse rate lower than that reported for these MS drugs. I am not trying to draw a conclusive statement about the effectiveness of LDN against other therapies based on this comparison, because the studies were not done in parallel. For instance, the ABC study was done only for RRMS, while my study includes patients with all types of MS. This could skew the average to a lower relapse rate, since relapses are not a prominent feature for patients in the progressive states of the disease. However, the RRMS subset of my survey was 116 subjects, or 68 percent of the total, and this group also reported a very low relapse rate of 0.26 per year.

The subjects in my survey were self-selected, meaning that they volunteered to participate rather than being randomly selected, another reason I do not construe this as a scientific study. But it would seem that this sort of positive flag from a sizeable group makes a good epidemiologic argument that larger human trials are warranted to establish the effectiveness of LDN in treating MS. The length of time the subjects had remained with the treatment is another indication of its effectiveness; the average duration was eight months, and 24 percent—64 of them—had been using it for two or more years.

Furthermore, as Dr. Agrawal points out, this patient survey is valuable because it indicates that LDN can make a positive difference in a disease like MS for which there are limited effective treatments, especially when the available drugs carry such a high price tag in terms of both economic cost and side effects. It also confirmed to the 267 patients in the survey group what we already knew: LDN was helping us.

Survey Population:

- 267 subjects, average ten years diagnosis, 65 percent female
- Average LDN treatment eight months, 24 percent 2 years+ of LDN treatment
- 10 percent, 28 individuals out of 267, reported a total of 42 relapses, 0.2 /year

Survey Results:

TABLE 9.1 SAMMYJO'S INTERNATIONAL PATIENT SURVEY

Type of MS	PPMS	PRMS	RRMS	SPMS	Total
	13%	4%	43%	39%	267
Months on LDN (Average)	10	13	7	9	8
Relapse rate	0.07	0.23	0.26	0.25	0.2
Subjective Assessments:					
Symptom improvement	53%	75%	82%	57%	70%
Progression halt	50%	58%	34%	43%	45%
LDN helpful, will continue	76%	83%	75%	70%	76%

Skip's Pharmacy LDN Survey

Skip Lenz, the compounding pharmacist of Skip's Pharmacy in Boca Raton, Florida, is a major supplier of compounded LDN and is in close contact with many of his clients who are prescribed LDN. With the help of pharmacy student interns, Dr. Lenz has created a patient survey on LDN and presented his results at the annual LDN conferences. The survey information is also available on his Web site, www.skipspharmacy.com. Lenz also publishes an ongoing blog about the latest LDN news.

His survey design is important to note. Although the patient responses are still considered anecdotal, Dr. Lenz utilized a randomization technique across all the LDN prescriptions he has filled (around 6,600 going back to 1999) along with an additional 1,200 prescription records collected from other local pharmacies, for a total record set of 7,800. From these records, patients were randomly selected. Randomization ensures that subjects will be included who had unfavorable as well as favorable results with LDN.

2006 Study

Dr. Lenz employed pharmacy candidate interns from the University of Florida to collect the survey data from this subset via patient phone interviews. The patient experience occurred over a time period spanning December 1, 2000 to June 30, 2005. The number of patients surveyed was 242 (p=v0.5); of these, 207 patients had MS.

At the 2006 LDN Conference, the survey presented showed patients' assessment of the effectiveness of LDN for treating their MS:

- improved   107   (52%)
- no change   86   (42%)
- worsened   12   (6%)
- mixed results   2   (1%)

TABLE 9.2 DR. LENZ'S 2006 PHARMACY PATIENT SURVEY

Type of MS	PPMS	CPMS	RRMS	SPMS	Unknown	Total
Number of patients	13	8	79	35	72	207
	(6%)	(4%)	(38%)	(17%)	(35%)	207
Subjective Assessments:						
Improved	10	1	48	17	31	107
	(77%)	(13%)	(61%)	(49%)	(43%)	(52%)
No Change	3	6	27	17	33	86
	(23%)	(74%)	(34%)	(49%)	(46%)	(42%)
Worsened			3	1	8	12
			(4%)	(2%)	(11%)	(6%)
Mixed		1	1			2
	(13%)	(1%)				(1%)

2006 Survey Presentation—
Relapses and Side Effects of LDN for MS

At the LDN conference in 2007, Dr. Lenz reported more results from the ongoing survey project, which has now grown to 278 patients randomized from the client list. The patient's definition of what constitutes an exacerbation was relied upon, since the survey did not involve physician records. The data are very promising and show that:

- 83 percent of those with MS reported no exacerbation
- no patient reported more than 1 relapse
- 71 percent felt their life had been improved by LDN
- 74 percent had no side effects (across all types of MS)
- 20 percent had one side effect (primarily sleep disturbance)
- 6 percent had two side effects (sleep disturbance plus stiffness or other minor side effect).

The First LDN Conference

The first LDN conference was held in New York City in the spring of 2005. The conferences have since continued annually, and the presentations are available online at ldninfo.org. Bernard Bihari's partner, Dr. David Gluck, organized the first conference, and he invited me to speak on an advocates panel and present the results of my survey. I was not aware that Skip's Pharmacy was conducting an LDN patient survey as well, so it was interesting to see that both surveys found similar positive results in MS.

Scientists interested in LDN research came from around the world, several researchers presented papers, and practicing physicians described the success they were seeing with LDN in their patients. Dr. Mir of the Evers Clinic in Germany presented his trial of LDN for MS. Before the conference I got a preview when he asked me to assist him with the English translation of his report. Dr. Maira Gironi from Italy described her observations of abnormally low endorphins levels in MS patients. She would later go on to conduct a trial of LDN's effects on primary progressive MS.

Many of the eighty attendees were friends I'd developed in the LDN online community. Robert Lester, a volunteer with the Accelerated Cure Project for MS (ACP) reported on his interview with Yash Agrawal following publication of Dr. Agrawal's hypothesis.

Patients Fund LDN Research

At the LDN conference Rob Lester introduced me to another presenter, Art Mellor, CEO of ACP. We formed a committee, with Dr. Agrawal as our physician advisor, and set up the LDN for MS Research Fund. The original goal was to raise $20,000 to fund basic laboratory research, the animal studies required before human trials. The committee members made initial donations to kick things off. Doctors Bihari and Gluck published an endorsement of our effort, contributed their own funds and offered to contribute the attendance fees from the second LDN conference. Skip Lenz at Skip's Pharmacy enclosed a message about our fund, with all their LDN prescription receipts urging customers to donate. We all felt passionately about the fund, and we knew that those who had benefited from this drug would too. So we reached out to the LDN community.

The "LDNers" who had the means to contribute responded to our call. Funds poured in via online contributions from visitors to my Web site. As the word spread, spontaneous fund-raising events broke out around the country. In Colorado, Shannon Doherty's mother had found relief from MS with LDN. To help our efforts, Shannon collected pledges and hiked the Colorado Trail. Bill Roberts, a former firefighter who got out of his MS wheelchair with LDN, joined the cause and "passed the fireman's boot" in Alabama for donations.

Around this time, I was contacted by an MS patient from California, Vicki Finlayson. LDN had stopped her severe neuropathic pain. An LDN advocate, she was on a mission to share her story and help others. She organized a gala benefit at her country club in Auburn, California. Here, Vicki and I told our LDN stories to the gathering of over 200. Mary Anne Bradley, author of *Up the Creek with a Paddle* came from the East Coast, and spoke about LDN helping stabilize her husband's progressive MS. She also described LDN improving her uncle's Parkinson's disease, and helping her mother battle cancer.

In one evening, the benefit Vicki organized raised the majority of our total contributions and we were able to reach our goal of $25,000. Now all we needed was a researcher interested in running a clinical trial. When I returned home, I was contacted by a gentleman who had attended Vicki's fundraiser. He had a tip for us. Since starting LDN, he had regained the ability to ride his bike long distance and achieved remission. (He's still at it; he logged 3000 miles in 2007.) This person told me that his neurologist, Dr. Bruce Cree, from the MS Center at the University of California, San Francisco (UCSF), had mentioned that he was considering a human study of LDN for MS.

The committee knew that this would be leaping ahead of standard scientific practice, but we also knew that the National MS Society had awarded Dr. Zagon at Penn State funding for an animal study on LDN and MS. We felt it would help our patient awareness effort if a human trial started as well. Lack of a human study is the biggest problem patients face when seeking a prescription for a new treatment. Where's the science? If doctors could see a trial underway at UCSF, we decided, this would help LDN's credibility.

I called and spoke to Dr. Cree at UCSF. He said that with the addition of our funds, the study could start immediately. Dr. Cree designed a double-blind placebo-controlled trial crossover study for eighty subjects, using a simple quality-of-life questionnaire to track results. The primary goal was to identify sufficient positive effects to justify a larger multiyear study that would include the full slate of tests needed to fully investigate the effects of LDN. The trial results are included in chapter 3.

Follow-up

Although I'm very happy and still amazed by my improvement, I don't intend to make a sales pitch for LDN. Its true efficacy needs to be proven scientifically with clinical trials.

Early pilot studies and phase 2 trials of LDN are already showing promise in conditions such as MS and Crohn's disease, as well as cancer. My goal and that of Vicki Finlayson, Mary Bradley and other patient advocates for LDN, who so willingly continue to share their experiences with LDN, is to encourage funding for full-scale clinical trials. As of May 2008, over ten thousand people have signed the online petition titled "Sign & Support the Campaign for Research & Trials in Low Dose Naltrexone for Multiple Sclerosis" available at http://www.thepetitionsite.com/takeaction/110785607.

Choosing LDN

Whether to try LDN for one's medical condition must be seriously considered, especially since it can have interactions with existing medical regimens, particularly if medications for pain are needed. Often, the choice is easiest for patients like me who have run out of other options.

Having failed to slow my disease progression with almost every FDA-approved medication available for MS, including a horrendous year of chemotherapy that left me with heart damage, I felt I was in the "last resort" category, and had nothing to lose. But what if I had known about LDN two years earlier, when I was making the decision to try chemotherapy? My neurologist had told me that chemotherapy was my only choice, even though risks included leukemia and heart damage. She said that I should never have children after taking this particular chemotherapy drug because it damages DNA. If I had known about LDN then, it would have been a no-brainer to try before agreeing to chemotherapy.

Fortunately, because of its low risk of side effects, LDN is regarded as falling in the "do no harm" category. Factors to take into consideration include available research studies and clinical trials of LDN in one's condition, strength of anecdotal reports, severity of your condition and rate of progression, drug contraindications, and one's response to their current therapies.

Working With Your Doctor

Today more and more doctors are familiar with LDN, and an increasing number of prescriptions are written for LDN each year. However, many doctors are still unfamiliar with LDN. Be prepared to discuss LDN with your doctor if it's new to him or her. See the resource section of this book for suggested publications and pharmacy resources that can help you acquaint your doctor with LDN. LDN involves a certain amount of risk, and in some cases, like mine, it means taking matters into your own hands and making the extra effort to consult with another physician.

About LDN

Naltrexone is a pure opiate antagonist that preferentially blocks the mu opiate receptor. Low doses of naltrexone are approximately 0.1mg/kg/day. Naltrexone is primarily eliminated by the kidneys, although a small amount is recycled by the liver.

Naltrexone is rabidly absorbed after it is ingested, primarily in the gastrointestinal tract. Peak plasma levels of naltrexone occur about one hour after the dose, and the effect of naltrexone lasts from twenty-four to seventy-two hours when doses of 50–300 mg daily are used. Lower doses of naltrexone (used in the LDN protocol) effectively block the opiate receptor for four to seven hours. The half-life (time when the original dose is reduced by 50 percent) of naltrexone is approximately three hours. Naltrexone is metabolized to 6-naltrexol, which is a weaker antagonist but has a longer half-life of approximately thirteen hours.

In the absence of opiate drugs, naltrexone causes few side effects, especially when low doses are used. Naltrexone can increase the secretion of gonadotropin-releasing hormone and corticotropin-releasing hormone and elevate the plasma concentrations of several hormones, including adrenocorticotrophic hormone (ACTH).

When to Try LDN

For patients not responding to conventional therapies or who have conditions for which no effective treatment is available, LDN may be a good option. For those newly diagnosed with conditions that have available therapies such as MS and Crohn's, it could help to select a drug regimen compatible with LDN (see the paragraph below headed "Combination Therapy"). Be aware that there may be higher risks of developing antibodies to your medication if you start and stop a drug like Remicade for Crohn's. In light of the promising pilot trials of LDN in Crohn's disease, it might be beneficial to try LDN first.

Research Summary

The specific diseases for which LDN and related opiate antagonists have been studied, or in which studies are currently underway, include pancreatic cancer, squamous cell carcinoma of the head and neck, HIV, Multiple Sclerosis, fibromyalgia, Crohn's, and Irritable Bowel Syndrome (IBS). See Appendix A for a complete listing with contact information.

The results of several clinical trials are being published as this goes to press. If you have one of the conditions in which the use of LDN is being investigated, clinical trial results may be available to help you with your decision. See the Appendix A for studies completed to date.

What If Your Disease
Has Not Been Studied?

If there are no clinical trial results available for your condition, review the studies available for conditions most similar to yours. Many chronic autoimmune conditions respond to the same medications. If there is a study indicating that LDN helps Crohn's disease, this will be of interest to any doctor who treats other autoimmune conditions, such as ulcerative colitis, lupus, rheumatoid arthritis, or autoimmune thyroid disease.

What to Expect from LDN Therapy

The length of time it takes to know if LDN is working varies widely, with improvement noted immediately to as long as nine months later. The primary goal of LDN therapy is slowing or stopping disease progression. In addition, symptoms may improve. Improvements seen in cancer include decreases in tumor size, disease remission, decreased tumor marker test results, and life extension beyond the expected norm for the particular condition. Improvements recorded in the pilot Crohn's disease clinical trial included decrease in disease index scores, symptom improvement, and incidence of disease remission.

Increased Endorphins

LDN causes an increase in endorphins that results in a feel-good effect. Mood elevation has been noted in the human trials of LDN, especially in studies where quality of life scores are monitored. Feeling good can alleviate stress, reduce depression, and induce healing. This is especially true for conditions like MS, where stress can lead to exacerbations.

Rehabilitation

Those with a neurodegenerative or physically wasting condition should check with their doctor about participating in a physical rehabilitation program and a nutritional support program. LDN may have stopped my disease progression, but it could not restore the physical abilities and strength that I'd lost. As you start to experience improvements with LDN there are two important concepts to restoring muscle tone and balance, "Don't overdo it" and "Use it or lose it."

When I started to feel better, I felt like I had a new lease on life, and so much to catch up on. I would overextend myself, because I didn't yet know what my new limits were. While exercise is needed to restore muscle tone, get adequate rest and allow time to heal.

After nine years of battling MS, I was in a state of physical decline, with atrophied muscles, and nerves that had lost contact with the muscles. As my energy increased I was able to start a physical therapy program. The following therapies have been found useful for individuals recovering from neurodegenerative diseases (the Resources has information on locating a practitioner in your area):

- physical therapy
- massage
- yoga
- craniosacral therapy
- Feldenkrais therapy and exercises
- chiropractic care

Safety

As described in chapter 1, naltrexone was initially tested for safety at the 50–100 mg dose level. Newer studies such as the Crohn's study and the Italian PPMS study have assessed naltrexone administered at low dose for safety and found no issues to date. In addition there are other mu opioid receptor antagonists, similar to LDN, under review by the FDA for the treatment of gastrointestinal conditions. These studies can also be used to assess the safety of LDN. For instance, Wyeth Research is seeking approval of methylnaltrexone in low dose format to treat constipation caused by opioids (see appendix).

Side Effects

The doctors we consulted for this book feel that LDN is harmless at the low doses used in low dose protocols. In studies of LDN there have been no reports of liver toxicity. However, because of the potential for liver toxicity to occur when high doses are used (50–300 mg naltrexone daily), and because of the metabolic changes that can occur in patients with liver disease, patients with preexisting liver or renal conditions using LDN should have their metabolic functions monitored by their doctors.

Another concern is the lack of research regarding the long-term effects of intermittent opiate blockade by LDN. Naltrexone has different effects when used in high doses (see chapter 7), and it's unknown whether the long-term use of LDN could have effects similar to those of high dose naltrexone (HDN). Consequently, LDN should be approached with caution by those who do not have a serious disease condition or are thinking of taking it as a preventative medication until long-term safety trials have been conducted.

Contraindications

LDN can be taken with other medications or supplements as long as they do not contain opiates or synthetic narcotics, examples of which include meperidine, fentanyl, tramadol (Ultram), morphine, Oxy-Contin, Vicodin, Percoset). Naltrexone blocks the body's opiate receptors. Therefore pain medications will be blocked from working and could lead to withdrawal problems. Check with both your doctor and pharmacist to make sure that none of your medications are contraindicated. They can also advise you on stopping pain medications that might interfere with LDN, and offer advice on the amount of time to allow between stopping opiates and starting LDN.

After starting LDN, if you have surgery scheduled or a procedure that may require pain medications, consult with your doctor to determine the amount of time needed to clear LDN from your system so that it does not interfere with anesthesia or pain medications. LDN must also be stopped if your doctor plans to prescribe opiate-based medications for post-operative use. The time required to clear naltrexone from the body may vary, based on dosage and body weight. Ian Zagon's animal studies revealed that a dosage of 0.1 mg/kg produced an opiate blockade that lasted for six to eight hours; this is slightly higher than the standard human dose of 4.5 mg/60 kg.

Combination Therapy

LDN upregulates the immune system and helps restore homeostasis. LDN doesn't stimulate the immune system. However, it's recommended that LDN not be used with immunosuppressant drugs. For MS, LDN is considered incompatible with the interferons (Beta Seron, Avonex, or Rebif). No adverse reactions have been reported when using combinations of these "incompatible" therapies, but they might cancel each other's effectiveness. Note that if you are on Copaxone for MS, this is not an issue, as it is not immune suppressive and is considered compatible with LDN.

Side Effects

Any potential side effects should diminish as the body adjusts to LDN and increased endorphin levels. Side effects are less likely to occur when a small starting dose is used and gradually increased. Experts advise to always err on the side of caution and reduce the dose if you notice an increase in symptoms.

Sleep

Sleep disturbances, including insomnia and vivid dreams, are the most commonly reported side effects associated with LDN. Sleep disturbances can lead to fatigue the following day. Dr. Zagon recommends taking LDN in the morning if it interferes with sleep when it's used at night.

Stiffness

In cases of MS, LDN may cause an initial increase in MS symptoms during the first week or two, especially stiffness, spasticity or muscle spasm. Stiffness is also a common symptom with Parkinson's. Dr. Zagon recommends reducing the dose if muscle stiffness occurs as a new symptom or if it increases.

Compounding and Dosage

Because naltrexone is FDA-approved and available in doses of 50–300 mg, prescriptions for low dose naltrexone must be prepared by a specialized compounding pharmacy. Alternately, you can receive a prescription for 50-mg tablets and dissolve and dilute the tablets to make the prescribed low dose.

Because of differences in compounding pharmacies and the fillers, it's suggested that patients use one of the compounding pharmacies listed in the Resources section, which includes pharmacies known to have experience with LDN. Most of the compounding pharmacies listed operate on a mail-order basis, and the average cost of LDN with shipping in the United States is less than $30 per month. Some American pharmacies also ship LDN internationally. There are also LDN Internet user groups in several languages that provide information on compounding pharmacies available in other countries. In addition, the U.S. LDN user groups maintain lists of U.S. and international pharmacies, along with a list of prescribing doctors.

To locate a local compounding pharmacy, call pharmacies in your area and ask if they're familiar with low dose naltrexone. Also, have them specify what type of filler they use. If they're unfamiliar with LDN, use fillers that aren't recommended, or indicate that they grind up the 50-mg tablets to put in the capsule, call another pharmacy. If possible, select a pharmacy before you see your doctor so he or she will know where to fax the prescription or who to consult if questions arise.

Fillers

For LDN, the pharmacy must *not* use a timed-release filler or anything that would make the drug release slowly into your system. The type of filler can also be important because of food sensitivities.

Lactose is commonly used as a filler, but it can cause upset for those with lactose intolerance or milk allergies. I prefer using a brand name filler called Avicel®, composed of a plant cellulose material. Questions have been raised about calcium carbonate filler, since it may pack too tightly, slowing the drug's release. If reactions to fillers are suspected, your pharmacist can prepare a liquid suspension of LDN. A transdermal patch or cream is sometimes prescribed for autistic children, although these preparations may result in a longer duration of absorption than is recommended for LDN. Oral preparations are recommended because their absorption is more predictable. Discuss your prescription with your doctor, and ask him or her to specify the filler.

When refilling your prescription, verify that the pharmacy is using the same filler you've used. Some patients on LDN report that they experienced a change in effectiveness when their formulation was changed without their knowledge.

Preparing Liquid LDN

Compounded capsules are the preferred method for ease of use. For patients allergic to capsule fillers, a liquid preparation may be worth discussing with your doctor and pharmacist. Or patients may prefer diluting their own LDN, which is generally less expensive. Here is a suggested method for making a liquid-based preparation from tablets, using the supplies that can be found at most pharmacies:

Supplies

- prescription of 50-mg naltrexone tablets
- 5-ml syringe or graduated baby medicine dropper
- 4-oz amber glass jar with a tight-fitting lid
- distilled water

Method

1. Using the syringe or dropper, add 5 ml of distilled water to the 4-oz jar.
2. Repeat ten times to get 50 ml total.
3. Mark the jar at the 50-ml level (with a permanent marker) for next time.
4. Add the tablet or, if it's a capsule, open and release into the water.
5. Shake until dissolved, and each time before taking a dose.
6. Use the syringe or dropper to measure the desired dose.
7. Remember that 1 ml contains 1 mg (1 mg/ml); e.g. for a 3-mg dose, use 3 ml.
8. Keep refrigerated and out of direct light; it is stable for thirty days.

Bihari's Protocol

In his initial research with LDN, Dr. Bihari prescribed doses of 1.5 to 4.5 mg taken at bedtime. The bedtime dosing schedule evolved from Bihari's theories on LDN causing a peak endorphin increase during the predawn hours.

However, studies show that taking LDN at night is not necessary. In the protocol he adhered to, Bihari didn't prescribe amounts greater than 4.5 mg and he didn't take body weight into account. He recommended lowering the dose if side effects occurred and emphasized that in MS patients with spasticity, a 3.0-mg daily dose of LDN was generally better tolerated than a 4.5-mg dose.

Body Weight Protocol

Ian Zagon recommends using a dose ranging from 3–10 mg daily. He has suggested starting out on a dose of 4.5 mg per 60 kg (150 lbs), adjusted to the patient's weight. This is based on the dose used in animal research at Penn State University, extrapolated from mouse body weight to human. This formula serves as the target dose, but this can vary, with some patients doing best taking LDN every other day or every third day. Patient experience and trial results have shown that effectiveness and minor side effects are surprisingly responsive to small changes of 1–2 mg per dose. In addition, initial adjustment to the drug, especially its impact on sleeping patterns, is facilitated by a gradual titration schedule.

Body Weight	1st Month	2nd Month	Ongoing
100	1 mg	2 mg	3 mg
150	1.5 mg	3 mg	4.5 mg
200	2 mg	4 mg	6 mg
250	2.5 mg	5 mg	7.5 mg

Three-month schedule by body weight in pounds, total daily milligram dose

An initial prescription of 180 compounded 1.5-mg capsules is sufficient to cover this protocol for most people. Here, a person weighing 150 pounds starts out using one capsule and adds another capsule the second and third months until the target dose is achieved. Once the optimal dose is determined, it can be ordered as a single capsule.

The most common approach in practice is to stop dose increases at 4.5 mg. This table is included as a suggestion for patients of varying weights, so they have dose options to explore if they are not achieving expected results. Patient experience has shown that the best approach is not to follow a one-size-fits-all schedule, rather to keep a log of dose, response and side effects, gradually adjusting the dose based on results.

What If You Cannot Find a Doctor Who Prescribes LDN?

If you cannot find a doctor in your area familiar with LDN, you may wish to consult by phone with a physician with experience prescribing LDN. You can ask members of the various LDN discussion forums for referrals in your area. You can find a list of LDN forums for various diseases in the resources section at LDNers.org. The following doctors provide telephone consultations and have agreed to have their contact information published. Contact their offices for information on fees and to learn what medical records you may need to fax them to establish your diagnosis.

Dr. John M. Sullivan
1001 S. Market St., Suite B, Mechanicsburg, PA 17055
Phone: 717-697-5050
Fax: 717-591-0920

Ardis Fisch, M.D.
192 Main St., Lee, MA 01238
Phone: 413-822-2096
Fax: 413-243-8227

Jeffrey Dach, M.D.
4700 Sheridan, Suite T., Hollywood, FL 33021
Phone: 954-983-1443

Dr. Steven G. Ayre
322 Burr Ridge Parkway
Burr Ridge, IL 60527
Phone: 630-321-9010

Pharmacists can also be helpful when it comes to finding doctors in your area who prescribe LDN.

10

The Potential Benefits and Future of LDN

Animal studies, clinical trials in humans, and case histories to date suggest that, used in low doses, the opiate antagonist naltrexone can stop disease progression and offer benefits in a number of autoimmune, infectious, and neurodegenerative conditions and in several different types of cancer. A wealth of anecdotal evidence, much of which is available online, suggests that low dose naltrexone (LDN) may benefit a broad range of disorders, essentially any condition that fits into the disease categories listed above.

Low dose naltrexone is reported to primarily exert its effects by increasing levels of endogenous opioid compounds. In particular, the increased production of [Met5]-enkephalin (opioid growth factor) and its receptor account for the most significant effects associated with LDN. By virtue of these effects, LDN inhibits cell proliferation, reduces inflammation, and promotes homeostasis, allowing the body to engage its systems toward the restoration of health. In addition, low dose naltrexone is suspected of having properties of its own that foster healing and it increases feelings of well-being related to increased endorphin production.

This chapter describes the various theories that have been proposed to explain how low dose naltrexone works on cellular and biochemical levels. In addition it describes the current status of LDN as a medical therapy, global LDN initiatives, the role of drug policies, contributions of the Internet, and expectations for LDN in the future.

Low Dose Naltrexone

Ian Zagon explains that low dose naltrexone is a misnomer in that the protocol called LDN refers to therapies involving low doses of any of the opiate antagonists (naloxone, naltrexone, nalmefene and others) or opiate growth factor (OGF), and also their metabolites and analogs.

In the LDN protocol, opiate antagonists are taken once daily in low doses.

For naltrexone, the recommended doses are 3–10 mg daily depending on body weight and other factors. Because doses are based on body weight, small children may use doses as low as 1 mg daily.

OGF (metenkephalin), which is primarily used in cancer and in early studies of HIV infection (for Bihari and Plotnikoff's study of metenkephalin in patients with AIDS-related complex, see chapter 8), has to be administered at regular intervals. In cancer, different dosing schedules for OGF are customarily used, depending on the estimated "tumor burden" present in the body.

Ian Zagon

Ian Zagon is the country's leading researcher when it comes to the use of opiate antagonists in conditions other than opiate and alcohol abuse. Zagon emphasizes that the immune system itself may not play much of a role in the effects attributed to LDN in cancer and other conditions. For example, LDN can inhibit tumor cell growth in tissue cultures, a scenario in which immune system cells do not exist. LDN can also inhibit tumor cell growth in nude mice, which have deficient immune systems. In studies where Zagon and his team have knocked out the opioid growth factor (OGF) receptor (OGFr), LDN has no effect. In experiments where they've introduced more OGF and OGF receptor, tumor inhibition is more robust. Zagon has proven that the effects of LDN in cancer are related to the system or complex formed by OGF and its receptor, and the role this system plays in cell proliferation.

However, increases in NK cells and other immune system components related to increased endorphins suggest that the increased endorphin levels resulting from LDN have significant effects on immune function. Consequently, LDN has the ability to restore homeostasis.

Endorphins

Endorphins are natural endogenous opiates produced by the adrenal gland, the pituitary gland, and the hypothalamus. Endorphins are neurotransmitters as well as cytokines and are important regulators of the immune response. Endorphins have effects commonly called the "runner's high" because of their analgesic and euphoric properties, which cause a feeling of well-being.

Four distinct groups of endorphins have been identified, including alpha-endorphin; beta-endorphin, gamma-endorphin, and sigma-endorphin. These different types of endorphins, like other polypeptide hormones, are synthesized in a "prepro" form that is one gigantic polypeptide that can mature into any of the endorphins. For example, the pituitary multi-hormone precursor termed proopiomelanocortin (POMC) contains the sequences for beta-lipotropin, melanocyte-stimulating hormone (MSH), endorphins, enkephalins, and adrenocorticotropic hormone (ACTH). After synthesis, this peptide is cleaved in the pituitary to generate ACTH and beta-lipotropin, while further processing within the central nervous system produces endorphins and enkephalins.

In an interesting study conducted by Maira Gironi and her team at the University of Milan, researchers found that beta-endorphin levels in patients with MS were significantly lower than those found in healthy controls. In addition, among the subtypes of MS, the lowest beta-endorphin levels were seen in patients with the more progressive forms of MS. The highest concentrations were found in patients with relapsing remitting, benign and clinically relapsing forms of MS.

Study results suggest that beta-endorphins may play a protective role in restoring cytokine balance. Endorphins are known to shift the Th1, Th2 balance toward Th2, which is mediated by decreased production of interleukin 12 (Il-12) by macrophages. This has been demonstrated in patients using LDN therapy.[1]

Opioid Growth Factor

The production of [Met-5]-enkephalin, a metenkephalin also known as opioid growth factor (OGF), increases in response to the opiate receptor blockade caused by LDN. Increased production of OGF is thought to be responsible for the majority of beneficial effects associated with LDN.

Cell Growth

Opiate antagonists can both inhibit and accelerate cell growth depending on the dose used and the duration in which the opiate receptor is blocked. With constant blockade, cell growth is stimulated. In intermittent blockade, cell growth is inhibited. LDN inhibits cell growth by causing an intermittent receptor blockade, making it a particularly effective treatment for cancer and infection.

Homeostasis

Homeostasis is a mechanism in which various bodily systems work together to maintain health. An example is the tolerance that occurs when the body is exposed to excessive doses of opiates. Pain itself causes the release of enkephalins. This is a homeostatic mechanism intended to help the body handle pain.

By binding to enkephalin receptors, opiates such as morphine enhance the pain-killing effects of enkephalin neurons. Release of enkephalins suppresses the transmission of pain signals. Through another homeostatic response, the sensitivity of the neural system to opiates decreases with continued use. In addition, when the drug is stopped, the system is no longer as sensitive as it initially was to the soothing effects of the enkephalin neurons. Instead, the pain of withdrawal is produced.

Oxidative Damage

Oxidative damage is caused by the production of reactive oxygen and nitrogen species. Normally, the body has sufficient antioxidant stores to prevent oxidative damage. In certain diseases, such as multiple sclerosis, reactive oxygen species such as superoxide, nitric oxide and peroxynitrite are dramatically increased and contribute to microglial activation. Microglia are immune system cells that reside in the central nervous system. Microglial activation is the hallmark of brain inflammation.

In various cancers, autoimmune and neurodegenerative diseases, the body's antioxidant resources are depleted by disease. This leads to increased levels of unresolved oxidative stress. Besides causing microglial activation, unresolved oxidative stress can damage the lipids, proteins and nucleic acids of cells and their energy-storing structures, the mitochondria. Mitochondrial DNA is particularly vulnerable to the effects of oxidative stress.[2]

By reducing the activity of enzymes that lead to free radical production, LDN reduces oxidative damage, including neuronal degeneration and demyelination. In doing so, LDN is able to halt disease progression in neurodegenerative conditions such as MS and Parkinson's disease.

Cell Signaling and Cytokines

The immune system and the brain communicate through signaling pathways. The brain and the immune system are the two major adaptive systems of the body. During an immune response this communication between the brain and immune system is essential for maintaining homeostasis. The main pathways of this communication are the hypothalamic-pituitary-adrenal axis (HPA axis) and the sympathetic nervous system.

The immune and nervous system cells release molecules known as cytokines that help modulate the immune response. Pro-inflammatory cytokines include interleukin-1 (IL-1), interleukin-6 (IL-6), interleukin-10 (IL-10), interleukin-12 (IL-1), interferon-gamma (IFN-gamma) and tumor necrosis factor alpha (TNF-α). Immune-system cells known as macrophages, which move quickly to the affected site during injury or infection, create cytokines and also microglia and astrocytes (immune cells of the nervous system), which trigger a sickness response.

Cytokines help modulate the immune response by sending signals to other cells. For instance, when macrophages present foreign antigens to T cells, cytokines are released that send signals to B cells to multiply. The B cells, in turn, send signals that result in antibody production. Cytokines can also promote inflammation and cause cytotoxic effects that contribute to cellular damage if allowed to persist. LDN induces production of endorphins that help modulate or regulate the immune response, reducing inflammation and associated cell damage.

Bihari's Studies

Neurologist Bernard Bihari is credited with initiating the first clinical studies of LDN. Before doing so, Bihari reviewed the studies of Ian Zagon and theorized how Zagon's findings might apply in a clinical setting. Because the early studies of naltrexone in drug abuse had already established the safety of this drug in higher doses, and because the use of naltrexone was already approved by the FDA as a treatment for drug abuse, Bihari was able to prescribe LDN off-label for other conditions.

The Immune System Boost?

In his early reports, Bihari wrote that LDN boosted the immune system. Studies have since shown that the effects of LDN lead to modulation rather than stimulation of the immune system. Many researchers describe LDN's ability to increase endorphin production. Endorphins are known to modulate or regulate the immune response.

In his book on psychoneuroimmunology, Jorge Daruna of Tulane University writes that beta-endorphin and other opioid peptides appear capable of suppressing or enhancing immune responsiveness. "For instance," he says, "antibody production in response to antigen stimulation can be modulated in a baseline-dependent manner. Beta-endorphin increased antibody production when baseline response was low and decreased it when it was high. Thus, opioid peptides have been regarded as fine-tuning immunomodulators."[3]

The Internet

With the help of his childhood friend Dr. David Gluck and Gluck's son, Bihari was able to share the results of his clinical success with LDN on an informational Web site. Since its inception, the Internet has been a popular vehicle for the spread of medical information. Researchers, physicians, students and patients all rely on the Internet to read about the latest medical findings. Thus, numerous patients, and to a lesser extent physicians and researchers, were able to learn of Bihari's reported success with LDN in a variety of different conditions. Even without clinical trials or peer-reviewed studies to confirm his findings, these anecdotal reports seemed destined to garner attention.

Each year found more and more people begin using the Internet. Consequently, once Bihari's Web site was introduced, an increasing number of patients began using LDN, either via consultations with Bihari or by finding local physicians who would prescribe it. Some of these patients started chat groups or Web sites to share their information, and many other patients participated on these boards or lurked online, absorbing every bit of information they could find. This Internet explosion was not confined to the United States. Web sites devoted to LDN began originating abroad, and the number of foreign researchers studying LDN also continued to grow.

The NCI Takes Notice

Health movements on the Internet are also noticed by government agencies and press agencies. It didn't take long for researchers at the National Center for Complementary and Alternative Medicine (NCCAM) division of the National Institutes of Health (NIH) to learn that Bernard Bihari was treating a significant number of cancer patients with LDN. Both NCCAM and the National Cancer Institute (NCI) were very interested in confirming the findings of Dr. Bihari and following up with studies of their own. NCCAM hired the Rand Corporation to study Bihari's files as part of a study for the Agency of Healthcare Quality and Research (AHQR).

Because many of Bihari's cancer patients were first seen and diagnosed by other physicians, his records were deemed incomplete. However, various departments of the NIH have been keeping their eye on the role of opiate antagonists. Most of the trials and studies in the United States on opiate antagonists have been partially or completely funded by NIH grants. In addition, the NIH sponsored a conference in April 2007, in which leading researchers, including Jill Smith and Ian Zagon from Penn State and Jau-Shyong Hong from the National Institute of Environmental Health Sciences (NIEHS) shared their research findings.

Clinical Trial Update

Clinical trials in the United States involving studies of opiate antagonists in conditions other than alcohol and drug abuse are steadily increasing, with trials occurring at Penn State, the University of California, San Francisco, and Stanford University. A larger number of trials have been conducted or are underway using opiate antagonists in moderate to high doses for smoking cessation, weight loss, drug abuse, kleptomania, pathological gambling, sex addiction (including Internet pornography addiction), bipolar disorders, eating disorders, and obsessive-compulsive disorders. These have taken place at Harvard, Yale, the University of Minnesota, and other universities in the United States, as well as universities in Tel Aviv, Cairo, London, and elsewhere abroad.

Safety and Efficacy

The safety of naltrexone and other opiate antagonists has primarily been studied at the higher doses used in addiction and compulsive disorders. In these studies, opiate antagonists have found to be generally safe. The safety of low dose naltrexone has been confirmed in trials of multiple sclerosis in Italy, and human trials of LDN in Crohn's disease and in animal studies of allergic autoimmune encephalomyelitis at Pennsylvania State University.

Drug Approval Policies

The current U.S. Food and Drug Administration (FDA) drug approval policy makes it difficult for LDN to be approved in medical conditions other than drug and alcohol abuse. For FDA approval for the new use of a previously approved drug, each specific use of LDN requires expensive clinical trials. The problems with approving off-label drugs are repeatedly discussed by critics of current FDA policies and cited as a hindrance to good medical care. In particular, the successful use of aspirin for reducing the mortality of heart attack victims was implemented for many years before the FDA finally approved this.[4]

Off-label Drugs

The sale of new drugs is forbidden unless these drugs have passed tests for safety and efficacy. This law is a provision of the 1962 amendment to the Food, Drugs, and Cosmetics Act of 1938. Physicians, however, can prescribe an FDA-approved drug for off-label uses. Overall, it's estimated that for each patient admitted to a U.S. hospital, at least one off-label drug is prescribed. In some cases, it's not unusual for a drug to be prescribed more often off-label than on-label. For instance, although thalidomide is approved for treating leprosy, it is more commonly used to treat conditions of multiple myeloma and AIDS.

The Economics of Prescription Drug Pricing

The off-label use of an inexpensive FDA-approved drug for medical conditions for which there are other, more expensive FDA-approved drugs available (the case when LDN is used in multiple sclerosis and AIDS) is frowned on by pharmaceutical companies and physicians who have ties to the drug industry. Pharmaceutical companies that have invested time and money developing new drugs stand to lose a great deal of money if the efficacy of LDN is proven in the medical conditions discussed in this text. Pharmaceutical companies have far greater ability to set prices in the United States than in other industrialized nations[5].

The increased amount of money spent on prescription drugs in the last decade is primarily related to the increased use of prescription drugs and, to a lesser extent, the increased cost of these drugs. Economists have found that increased drug use is related to increased prescription drug benefits available through both government and private insurance.

The low cost of generic naltrexone is not favorable to the profits of drug manufacturers who might otherwise implement clinical trials for its use in conditions for which expensive drugs are already available. Some pharmaceutical manufacturers have attempted to develop opiate-antagonist metabolites or analogs in an effort to produce drugs similar to LDN that would be different enough to patent and generate higher costs. Several biotech companies have shown interest in studying and manufacturing LDN, but to date none of these companies have initiated trials. An article by a Columbia University student in the *Columbia Spectator* describes the frustration of MS patients who understand the financial motives of pharmaceutical companies in keeping LDN out of the limelight.[6]

State Government Initiatives

In some states, patients with MS have joined with their local multiple sclerosis agencies to appeal to their elected officials to help generate funding for clinical trials in LDN.

Vermont

At the request of local MS Society members, in 2006 a bill was introduced by Representative Michael Obuchowski of Rockingham, Vermont, for state support to study LDN in the treatment of MS. The legislature resolved that the General Assembly urge both the FDA and the Multiple Sclerosis Society "to conduct scientifically valid clinical trials to assess the effectiveness and ramifications of low dose naltrexone as a medication for treating multiple sclerosis," that Congress appropriate funds to support the federal research, and that copies of this resolution be sent to Acting FDA Commissioner Dr. Lester Crawford, President and Chief Executive Officer of the National Multiple Sclerosis Society Michael Dugan, and the members of the Vermont Congressional Delegation.[7]

California

In March 2008, patients with MS appealed to the California state legislature to pass a bill similar to the bill proposed by citizens of Vermont.

Global Initiatives

LDN was chosen as the drug of choice for the Mali AIDS/HIV trial after health ministers in Africa studied all available evidence on naltrexone (for more information, see chapter 8). In Ireland, Mary Boyle Bradley's brother, Dr. Phil Boyle, is seeing excellent results using LDN in his practice. (Mary's story of procuring LDN for her husband is described in her 2005 book *Up the Creek with a Paddle*). In Yugoslovia, a team of researchers at the Immunology Research Center has shown how opiate antagonists involve various opioid receptor subtypes in immunomodulation.[8]

In Scotland, the LDN Research Trust, which was set up in England, has raised £9500 through charitable donations from patients to help pay for a trial and plans to support the Scottish LDN project. Dr Tom Gilhooly also undertook a twenty-eight-mile sponsored mountain-bike ride to raise extra money for the charity. He said that £50,000 was needed to fund the trial and that he intended to apply to Scotland's Chief Scientist Office for further assistance. Gilhooly said, "I have seen with my own eyes in my own practice over the last three years the significant improvements in people who really do not have much in the way of options."[9]

In a 1998 study in Croatia, researchers examined the effects of metenkephalin (Peptid-M) on human lymphocytes. Most importantly, they have shown that metenkephalin causes normalization of chromosomally aberrant cell findings in patients suffering from different immune-mediated diseases.[10]

The Future of LDN
in the United States

Dr. Jarred Younger, chief investigator for the LDN Clinical Trial in Fibromyalgia at Stanford University (see chapter 3), reports that Gulf War Syndrome (or Gulf War Illness) shares a number of overlapping symptoms with fibromyalgia and chronic fatigue syndrome. Gulf War Syndrome is a disorder seen in combat veterans of the 1991 Persian Gulf War, with symptoms of fatigue, joint and muscle pain, dermatologic conditions, and dyspepsia.[11] Additional symptoms such as night sweats, headaches, and problems with thinking and memory are also frequently reported.[12] Recent meta-analyses have confirmed the increased risk of pain[13] and other symptoms[14] in individuals deployed to the 1991 Gulf War.

Gulf War Syndrome presents similarly to other multisymptom illnesses, which may be collectively referred to as central sensitivity syndromes.[15] Because of the possible overlap in pathophysiologic mechanisms, treatments successful in disorders such as fibromyalgia and chronic fatigue syndrome may also be helpful to those with Gulf War Syndrome. Dr. Younger plans to conduct a trial of LDN in Gulf War Syndrome in the near future.[16]

At Penn State, Ian Zagon continues his studies of LDN in pancreatic cancer and MS, and plans to study LDN in Parkinson's disease. He is particularly excited about his team's elegant studies that included the use of siRNAs to knock out the OGF receptor, and the cell kinase studies he has been conducting to demonstrate cell growth effects at the molecular level of perturbation. Zagon is currently conducting studies on head and neck cancer, pancreatic cancer, wound healing, and MS.

As Yash Agrawal says in the foreword to this book, thousands of people worldwide are experiencing benefits from LDN. Can so many people be wrong? However, it will take many more clinical trials before the true benefits of LDN are fully known.

Chapter Notes

Chapter 1

1. Julius Demetrios, "NIDA's Naltrexone Research Program," in *Narcotic Antagonists, Naltrexone: Progress Report*, ed. Julius Demetrios and Pierre Renault, NIDA Research Monograph 9 (Rockville, MD: U.S. Department of Health, Education and Welfare, 1976), 5.

2. Dan Wakefield, *New York in the 50s* (New York: Houghton Mifflin, 1992).

3. Jerome Jaffe, "Foreword," in Demetrios and Renault, vi.

4. Paracelsus wrote, "Alle Ding sind Gift und nichts ohn Gift; allein die Dosis macht das ein Ding kein Gift ist." (All things are poison and not without poison; only the dose makes a thing not a poison.)

5. Demetrios Julius, 8.

6. Monique Braude and J. Michael Morrison, "Preclinical Toxicity Studies of Naltrexone," in Demetrios and Renault, 16.

7. "The Corporate, Political, and Scientific History of Naltrexone," Gazorpa.com http://www.gazorpa.com/History.html (accessed May 13, 2008).

8. "Case Studies: LAAM, Naltrexone, Clozapine and Nicorette," U.S. Department of Health and Human Services, http://aspe.hhs.gov/health/reports/cocaine/4cases.htm (accessed May 13, 2008).

9. National Institute on Drug Abuse, http://www.drugabuse.gov (accessed July 30, 2007).

10. For more information, see U.S. Food and Drug Administration, Office of Orphan Products Development, http://www.fda.gov/orphan (accessed May 13, 2008).

11. Candace B. Pert, *Molecules of Emotion: The Science Behind Mind-Body Medicine* (New York: Simon & Schuster, 1999).

12. Louis Sanford Goodman, Alfred Gilman, Laurence L. Brunton, John S Lazo, and Keith L. Parker, eds., *Goodman & Gilman's The Pharmacological Basis of Therapeutics*, 11th edition (New York: McGraw-Hill, 2006) 547–549.

13. "'Off-Label' and Investigational Use of Marketed Drugs, Biologics, and Medical Devices," Guidance for Institutional Review Boards and Clinical Investigators, 1998 Update, U.S. Food and Drug Administration, http://www.fda.gov/oc/ohrt/irbs/offlabel.html (accessed May 13, 2008).

14. Samantha Wilkinson, "Low Dose Naltrexone for Multiple Sclerosis," http://www.ldners.org (accessed May 13, 2008).

Chapter 2

1. Jorge H. Daruna, *Introduction to Psychoneuroimmunology* (Boston: Elsevier Academic Press, 2004), 165.

2. Elaine A. Moore, *Autoimmune Diseases and Their Environmental Triggers* (Jefferson, NC: McFarland, 2002), 72–94.

3. B. A. Cree, O. Khan, D. Bourdette, S. Goodin, J. Cohen, A. Marrie, D. Glidden, et al., "Clinical Characteristics of African American vs Caucasian Americans with Multiple Sclerosis," *Neurology* 63 (2004), 2039–45.

4. Nikolaos Grigoriadis and Georgios Hadjigeorgious, "Virus-mediated Autoimmunity in Multiple Sclerosis," *Journal of Autoimmune Diseases* 3, no.1 (February 2006), http://www.jautoimdis.com/content/3/1/1 (accessed May 13, 2008); A. Chaudhuri and P. Behan, "Multiple Sclerosis Is Not an Autoimmune Disease," *Archives of Neurology* 61, no. 10 (October 2004), 1610–1612.

5. Russell Blaylock, *Excitotoxins: The Taste That Kills* (Sante Fe, NM: Health Press, 1994).

6. Y. P. Agrawal, "Low Dose Naltrexone Therapy in Multiple Sclerosis, *Medical Hypotheses* 64, no.4 (2005), 721–724.

7. Bin Liu and Jau-Shyong Hong, "Role of Microglia in Inflammation-Mediated Neurodegenerative Diseases: Mechanisms and Strategies for Therapeutic Intervention," *Journal of Pharmacology and Experimental Therapeutics* 304, no.1 (2003), 272.

8. "Cell Damage and Autoimmunity Colloquium Draws Internationally Known Researchers," *InFocus* (American Autoimmune Related Diseases Association) 16 no.1 (March

2008), 1–5; http://www.aarda.org/research_ display.php?ID=56 (accessed May 13, 2008). Excerpted from Ian Mackay, Natasha Leskovsek, and Noel Rose, "Cell Damage and Autoimmunity: A Critical Appraisal," *Journal of Autoimmunology* 30, no.1–2 (February–March 2008), 5–11.

9. Deling Yin, David Tuthill, R. Mufson, and Yufang Shi, "Chronic Restraint Stress Promotes Lymphocyte Apoptosis by Modulating CD9 Expression," *Journal of Experimental Medicine* 191, no.8 (April 2000), 1423–1428.

10. Daruna, *Introduction to Psychoneuroimmunology*, 170.

11. "MedInsight Announces Clinical Trial Results of Low-dose Naltrexone: Potential Breakthrough for Crohn's Disease Patients," *Medical News Today*, January 31, 2007, http://www.medicalnewstoday.com/articles/61960.php (accessed May 13, 2008).

12. Jill Smith, Heather Stock, Sandra Bingaman, David Mauger, Moshe Rogosnitzky, and Ian Zagon, "Low-Dose Naltrexone Therapy Improves Active Crohn's Disease," *American Journal of Gastroenterology* 102, no.4 (2007), 1–9.

13. Jill Smith, "Animal Studies with Naltrexone," presented at the NCI Conference, "Low Dose Opioid Blockers, Endorphins and Metenkephalins: "Promising Compounds for Unmet Medical Needs," Bethesda, Maryland, April 20, 2007.

14. David Keren and James Goeken, "Autoimmune Reactivity in Inflammatory Bowel Diseases," in *Progress and Controversies in Autoimmune Disease Testing*, ed. David F. Keren and Robert M. Nakamura (Philadelphia: W. B. Saunders, 1997), 465–481.

15. R. Kariv, E. Tiomny, R. Grenshpon, R. Dekel, G. Wasiman, Y. Ringel, and Z. Halpern, "Low-dose Naltrexone for the Treatment of Irritable Bowel Syndrome: A Pilot Study," *Digestive Diseases Sciences* 51, no. 12 (December 2006), 2128–2133; Joyce Generalia and Dennis Cada, "Naltrexone: Irritable Bowel Syndrome," *Hospital Pharmacy* 42, no. 8 (August 2007), 712–718.

16. "Positive Clinical Data in Irritable Bowel Syndrome to be Presented at Gastroenterology Meeting." Pain Therapeutics, Inc. press release, October 15, 2003.

17. "Effects of Low Dose Naltrexone in Fibromyalgia," Stanford School of Medicine Clinical Trials Directory, http://med.stanford.edu/clinicaltrials/detail.do?studyId=756 (accessed May 13, 2008).

18. Jaquelyn McCandless, *Children with Starving Brains: A Medical Treatment Guide for Autism Spectrum Disorder*, 3rd ed. (Putney, VT: Bramble Books, 2007). 242–243.

19. "LDN and Autoimmune Disease," http://www.lowdosenaltrexone.org/ldn_and_ai.htm.

20. Fred P. Sherman and David C. Atkinson, inventors, "Method of Treatment for Autoimmune Diseases," Free Patents Online, http://www.freepatentsonline.com/4857533.html (accessed May 13, 2008).

21. Sandi Lanford, "Low Dose Naltrexone and Autoimmune Disease," Revolution Health, http://www.revolutionhealth.com/blogs/nothing/low-dose-naltrexone-a-4313.

22. "LowDose Naltrexone: Treatment of Endorphin Deficiency," Fertility Care, http://www.fertilitycare.net/documents/LDNInfo_000.pdf (accessed May 13, 2008).

Chapter 3

1. "Who gets MS?" National Multiple Sclerosis Society, http://www.nationalmssociety.org/about-out-multiple-sclerosis/who-gets-ms/index.aspx (accessed May 13, 2008).

2. "The Cure Map," Accelerated Cure Project for Multiple Sclerosis, http://www.acceleratedcure.org/curemap/map.php (accessed May 13, 2008).

3. Kenneth Singleton, *The Lyme Disease Solution* (South Lake Tahoe, CA: BioMed Publishing Group, 2008); Singleton personal correspondence with author, February 22, 2008.

4. S. Goffette, V. van Pesch, J. L. Vanoverschelde, et al., "Severe Delayed Heart Failure in Three Multiple Sclerosis Patients Previously Treated with Mitoxantrone," *Journal of Neurology* 252 no. 10 (October 2005), 1217–22.

5. E. Le Page E, E. Leray, G. Taurin, et al., "Mitoxantrone as Induction Treatment in Aggressive Relapsing Remitting Multiple Sclerosis: Treatment Response Factors in a 5 Year Follow-up Observational Study of 100 Consecutive Patients," Journal of *Neurology, Neurosurgery & Psychiatry* 79, no.1 (January 2008), 52–6.

6. R. M. Ransohoff, "'Thinking without thinking" about Natalizumab and PML," *Journal of the Neurological Sciences* 21, no.2 (August 15, 2007), 50–2.

7. Patricia K. Coyle, "Management of Suboptimal Responders," in *MS Treatment: The Importance of Early Diagnosis and Comprehensive Disease Management*, ed. Robert M. Herndon; Douglas Goodin; Patricia K. Coyle; Amy Perrin Ross. Medscape Today, October 12, 2006; http://www.medscape.com/viewprogram/6005 (accessed May 13, 2008).

8. A. Chaudhuri and P. O. Behan, "Multiple Sclerosis: Looking beyond Autoimmunity, *Journal of the Royal Society of Medicine* 98, no. 7 (July 2005), 303–6.

9. "What Are You Using to Treat Your

MS? ThisIsMS, http://www.thisisms.com/survey-results-2—.html (accessed May 13, 2008).

10. "Can Naltrexone Relieve MS Symptoms," *New Horizons* (Brewer Science Library, Richland Center, Wisconsin), Winter 1999, http://www.mwt.net/~drbrewer/naltrexms.htm (accessed May 13, 2008).

11. Yash Agrawal, "The Need for Trials of Low Dose Naltrexone as a Possible Therapy for Multiple Sclerosis." Interviewed with Rob Lester of the Boston Cure Project, January 2005 (http://www.acceleratedcure.org/downloads/interview-agrawal.pdf, accessed May 13, 2008).

12. M. Gironi, R. Furlan, M. Rovaris, G. Comi, M. Filippi, A. Panerai, and P. Sacerdote, "Beta Endorphin Concentrations in PBMC of Patients with Different Clinical Phenotypes of Multiple Sclerosis," *Journal of Neurology, Neurosurgery & Psychiatry* 74, no. 4 (April 2003), 495–97.

13. Z. Mir, "Clinical Trial at the Klinik Dr. Evers, Study on the Symptomatical Effects of LDN in Multiple Sclerosis," Power Point Presentation, First Annual LDN Conference, New York City, April 2005.

14. Private conversation with Dr. Ian Zagon, May 29, 2008, June 1, 2008.

15. Martin L. Pall, "Chronic Fatigue Syndrome as a NO/ONOO-Cycle Disease," School of Molecular Biosciences publication, Washington State University, http://molecular.biosciences.wsu.edu/Faculty/pall/pall_cfs.htm.

16. Bruce Kornyeyva, "A Randomized Placebo-Controlled, Crossover–Design Study of the Effects of Low Dose Naltrexone," Trial NCT00501696; ClinicalTrials.gov, May 22, 2008 update, http://clinicaltrials.gov/ct2/results?term=NCT00501696.

17. Bruce Cree, Michael Ross, Ivo Violich, Brendan Berry, Azadeh Beheshtian, Elena Kornyeyeva, and Douglas Goodin, Department of Neurology, UCSF, San Francisco, CA, "A Single Center, Randomized, Placebo-Controlled, Double-Crossover Study of the Effects of Low Dose Naltrexone on Multiple Sclerosis Quality of Life," Presentation at the 60th Annual American Academy of Neurology Meeting, April 15, 2008.

18. Maira Gironi, Filippo Martinelli, Paola Sacerdote, Claudio Solaro (Genova, Italy), Rosella Cavarretta, Mauro Zaffaroni, Lucia Moiola, Marta Radaelli, Valentino Pilato (Gallarate, Italy), Sebastiano Bucello, Vittorio Martinelli, Marco Cursi, Raffaello Nemni, Giancarlo Comi, and Gianvito Martino (Milano, Italy), "A Pilot Trial of Low Dose Naltrexone in Primary Progressive Multiple Sclerosis," Presentation at the 60th Annual American Academy of Neurology Meeting, April 15, 2008.

Chapter 4

1. Bin Liu, H. M. Gao, J. Wang, G. Jeohn, C. Cooper, and Jau-Shyong Hong, "Role of Nitric Oxide in Inflammation-mediated Neurodegeneration, *Annals of the New York Academy of Science* 962, no.1 (May 2002), 318–33.

2. Yash Agrawal, "Low Dose Naltrexone Therapy in Multiple Sclerosis," *Medical Hypotheses* 64, no.4 (2005), 721–724.

3. Paul Aisen, Deborah Marin, and Kenneth Davis, "Inflammatory Processes—Anti-Inflammatory Therapy," in *Alzheimer Disease: From Molecular Biology to Therapy*, ed. R. Becker and E. Giacobini (Boston: Birkhauser Publishing, 1996), 349–353.

4. Bin Liu and Jau-Shyong Hong, "Role of Microglia in Inflammation-Mediated Neurodegenerative Diseases: Mechanisms and Strategies for Therapeutic Intervention," *Journal of Pharmacology and Experimental Therapeutics* 304, no.1 (2003), 1–7.

5. Bin Liu, Hui-Ming Gao, Jiz-Yuh Wang, Gwang-Ho Jeohn, Cynthia Cooper, and Jau-Shyong Hong, "Nitric Oxide Novel Actions, Deleterious Effects, and Clinical Potential, " *Annals of the New York Academy of Sciences* 962, no.1 (May 2002), 318–331.

6. Bin Liu and Jau-Shyong Hong, "Role of Microglia," 1–7.

7. Priti Patel, "Low-Dose Naltrexone for Treatment of Multiple Sclerosis: Clinical Trials are Needed," Letter to the Editor, *Annals of Pharmacotherapy*, September 2007.

8. S. Lipton, Z. Gu, and T. Nakamura, "Inflammatory Mediators Leading to Protein Misfolding and Uncompetitive/fast Off-rate Drug Therapy for Neurodegenerative Disorders, *International Reviews in Neurobiology* 82 (2007), 1–27.

9. Russell Blaylock, *Excitotoxins: The Taste that Kills* (Santa Fe, NM: Health Press, 1994), 30–31.

10. Wei Zhang, Jau-Shyong Hong, Hyoung-Chun Kim, Wanqin Zhang, and Michelle Block, "Morphinan Neuroprotection: New Insight into the Therapy of Neurodegeneration," *Critical Reviews in Neurobiology* 16, no.4 (2004), 271–302.

11. Yash Agrawal, "Low Dose Naltrexone Therapy," 721–724.

12. Gunther Deuschl, Carmen Schade-Brittinger, Paul Krack, et al, "A Randomized Trial of Deep-Brain Stimulation for Parkinson's Disease," *New England Journal of Medicine* 355 no. 9 (August 31, 2006), 896–908.

13. Bin Liu and Jau-Shyong Hong, "Role of Microglia," 1–7.

14. B. Liu and J. S. Hong, "Naloxone Protects Rat Dopaminergic Neurons against

Inflammatory Damage through Inhibition of Microglia Activation and Superoxide Generation," *Journal of Pharmacology and Experimental Therapeutics* 293, no.2 (May 2000), 607–617.

15. Yash Agrawal, "Low Dose Naltrexone Therapy,": 721–724.

16. Olivier Rascol, Nelly Fabre, Olivier Blin, et al., "Naltrexone, an Opiate Antagonist, Fails to Modify Motor Symptoms in Patients with Parkinson's Disease," *Movement Disorders* 9 no. 4 (October 2004), 437–440.

17. B. Liu, L. Du, and J. S. Hong, "Naloxone Protects Rat Dopaminergic Neurons against Inflammatory Damage through Inhibition of Microglia Activation and Superoxide Generation," *Journal of Pharmacology and Experimental Therapeutics* 293 no. 2 (May 2000), 607–617.

18. *LDN: The Latest News*, http://www.lowdosenaltrexone.org/ldn_latest_news.htm (accessed May 14, 2008).

19. Timothy Miller and Don Cleveland, "Treating Neurodegenerative Diseases with Antibiotics," Perspectives, *Science* 307 (January 21, 2005): 361–62.

20. Yash P. Agrawal, "Possible Importance of Antibiotics and Naltrexone in Neurodegenerative Disease," Letter to the Editor, *European Journal of Neurology* 13, no. 9 (2006), e7.

21. Tony E. White, personal correspondence with Samantha Jo Wilkinson, used with permission, April 29, 2007.

Chapter 5

1. Ian S. Zagon, *Opioids and Development: New Lessons From Old Problems, Prenatal Drug Exposure: Kinetics and Dynamics*, National Institute on Drug Abuse Research Monograph Series 60, United States Department of Health and Human Services, 1985.

2. A. Cornelius Celsus, *De Medicina*, Loeb Classical Library, vol. 3 (Cambridge, MA: Harvard University Press, 1935).

3. "The History of Cancer (Introduction)," Institut Jules Bordet, http://www.bordet.be/en/presentation/history/cancer_e/cancer1.htm (accessed May 14, 2008).

4. Sherri Stewart, Jessica King, Trevor Thompson, Carol Friedman, and Phyllis Wingo, "Cancer Mortality Surveillance—United States 1990–2000," MMWR Surveillance Summaries (Morbidity and Mortality Weekly Report), June 4, 2004, http://www.cdc.gov/mmwr/preview/mmwrhtml/ss5303a1.htm (accessed May 14, 2008).

5. "Targeted Cancer Therapies: Questions and Answers," National Cancer Institute, June 13, 2006; http://www.nci.nih.gov/cancer-topics/factsheet/Therapy/targeted (accessed May 14, 2008).

6. "Cancer and CAM," National Center for Complementary and Alternative Medicine, September 2005, updated June 2007, http://nccam.nih.gov/health/camcancer (accessed May 14, 2008).

7. "Understanding the Approval Process for New Cancer Treatments," National Cancer Institute, December 30, 1999, updated January 6, 2004, http://www.cancer.gov/clinicaltrials/learning/approval-process-for-cancer-drugs (accessed May 14, 2008).

8. "Faculty Biosketch: Ian S. Zagon, " Penn State College of Medicine, http://www.hmc.psu.edu/neuroscienceresearch/admin/facultybiopages/zagon.htm (accessed May 14, 2008).

9. P. J. McLaughlin, B. C. Stack Jr., R. J. Levin, F. Fedok, I. S. Zagon, "Defects in the Opioid Growth Factor Receptor in Human Squamous Cell Carcinoma of the Head and Neck," *Cancer* 197 no. 7 (April 2003), 1701–1710.

10. "Opioid Growth Factor Receptor—Homo sapiens (Human), Reviewed," Q9NZT2, UniProt Consortium, http://beta.uniprot.org/uniprot/Q9NZT2 (accessed May 14, 2008).

11. "Opioid Growth Factor," NCI Drug Dictionary, National Cancer Institute, http://www.cancer.gov/Templates/drugdictionary.aspx/?CdrID=428488 (accessed May 14, 2008).

12. I. S. Zagon, C. D. Roesner, M. F. Verderame, B. M. Olsson-Wilhelm, R. J. Levin, and P. J. McLaughlin, "Opioid Growth Factor Regulates the Cell Cycle of Human Neoplasias," *International Journal of Oncology* 17, no.5 (November 2000), 1053–1061.

13. Fan Cheng, Ian S. Zagon, Michael F. Verderame, and Patricia McLaughlin, "The Opioid Growth Factor (OGF)-OGF Receptor Axis Uses the p16 Pathway to Inhibit Head and Neck Cancer," *Cancer Research* 67 (November 1, 2007), 10511–10518.

14. Fan Cheng, Patricia McLaughlin, Michael Verderame, and Ian S. Zagon, "The OGF-OGFr Axis Utilizes the p21 Pathway to Restrict Progression of Human Pancreatic Cancer," *Molecular Cancer*, January 11, 2008, provisional abstract available online at http://www.molecular-cancer.com/content/7/1/5 (accessed May 14, 2008).

15. I. S. Zagon, K. A. Rahn, and P. J. McLaughlin, "Opioids and Migration, Chemotaxis, Invasion, and Adhesion of Human Cancer Cells, *Neuropeptides* 41, no. 6 (December 2007). 442–452.

16. J. Blebea, J. Mazo, T. Kihari, J. H. Vu, P. J. McLaughlin, R. G. Atnip, I. S. Zagon, "Opioid Growth Factor Modulates Angiogenesis," *Journal of Vascular Surgery* 32 no. 2 (August 2000), 364–373.

17. I. S. Zagon, M. F. Verderame, J. Hankins, and P. J. McLaughlin, "Overexpression of the Opioid Growth Factor Receptor Potentiates Growth Inhibition in Human Pancreatic Cancer," *International Journal of Oncology* 30 no. 4 (April 2007), 775–783.

18. J. P. Smith, R. I. Conter, S. I. Bingaman, H. A. Harvey, D. T. Mauger, M. Ahmad, L. M. Demers, W. B. Stanley, P. J. McLaughlin, and I. S. Zagon, "Treatment of Advanced Pancreatic Cancer with Opioid Growth Factor: Phase I," *Anticancer Drugs* 15 no. 3 (March 2004), 203–209.

19. "Excerpt from a S-1 SEC Filing, Filed by Innovive Pharmaceuticals," May 24, 2007, EDGAR Online, http://sec.edgar-online.com/2007/05/24/0000950144-07-005156/Section 13.asp (accessed May 14, 2008).

20. "Opioid Growth Factor in Treating Patients with Advanced Pancreatic Cancer That Cannot be Removed by Surgery," sponsored by Milton S. Hershey Medical Center, ClinicalTrials.gov, http://clinicaltrials.gov/show/NCT00109941 (accessed May 14, 2008).

21. Ian S. Zagon, Jeffrey R. Jaglowski, Michael F. Verderame, Jill P. Smith, Alphonse E. Leure-re-duPree, and Patricia J. McLaughlin, "Combination Chemotherapy with Gemcitabine and Biotherapy with Opioid Growth Factor (OGF) Enhances the Growth Inhibition of Pancreatic Adenocarcinoma," *Cancer Chemotherapy and Pharmacology* 5 no. 5 (November 2005), 510–520.

22. B. Berkson, D. Rubin, and A. Berkson, "The Long-Term Survival of a Patient with Pancreatic Cancer with Metastases to the Liver after Treatment with the α–Lipoic Acid/ Low Dose Naltrexone Treatment Protocol," *Integrative Cancer Therapies* 5 no. 1 (2006), 83–89.

23. P. J. McLaughlin, et al., "Defects in the Opioid Growth Factor Receptor," 1701–1710.

24. J. R. Jaglowski, I. S. Zagon, B. C. Stack Jr., M. F. Verderame, A. E. Leure-re-duPree, J. D. Manning, and P. J. McLaughlin, "Opioid Growth Factor Enhances Tumor Growth Inhibition and Increases the survival of Paclitaxel-treated Mice with Squamous Cell Carcinoma of the Head and Neck," *Cancer Chemotherapy and Pharmacology* 56 no. 1 (July 2005), 97–104.

25. B. Berkson, D. Rubin, and A. Berkson, "Reversal of Signs and Symptoms of a B-cell Lymphoma in a Patient Using Only Low-dose Naltrexone," *Integrative Cancer Therapies* 6 no. 3 (September 2007), 293–296, http://www.ldn4cancer.com/files/berkson-b-cell-lymphoma-paper.pdf (accessed May 14, 2008).

26. Authors' telephone conference with Ian Zagon and Judy Canfield, January 25, 2008.

27. Nicholas Plotnikoff, "Methionine Enkephalin, A New Cytokine with Antiviral and Anti-tumor Properties," chapter 19 in *Po-*

tentiating Health and the Crisis of the Immune System, ed. A. Mizrahi, Stephen Fulder, and Nimrod Sheinman (New York: Springer, 1997), 193–194.

28. I. S. Zagon and P. J. McLaughlin, "Naltrexone Modulates Tumor Response in Mice with Neuroblastoma," *Science* 221, no. 4611 (August 12, 1983), 671–673.

29. Ian Zagon and Patricia McLaughlin, "Opioids and Differentiation in Human Cancer Cells," *Neuropeptides* 39, no.5 (October, 2005), 495–505.

30. I. S. Zagon, S. D. Hytrek, and P. J. McLaughlin, "Opioid Growth Factor Tonically Inhibits Human Colon Cancer Cell Proliferation in Tissue Culture, *American Journal of Regulatory, Integrative and Comparative Physiology* 271, no. 30 (1996), 411–518.

31. Paolo Lissoni, Fabio Malugani, Ola Malysheva, Vladimir Kozlov, Moshe Laudon, Ario Conti, and Georges Maestroni, "Neuroimmunotherapy of Untreatable Metastatic Solid Tumors with Subcutaneous Low-dose Interleukin-2, Melatonin and Naltrexone, *Neuroendocrinology Letters* 23, no. 4 (August 200), 341–344, http://www.nel.edu/23_4/NEL 230402A09_Lissoni.htm (accessed May 14, 2008).

32. G. J. Bisignani, P. J. McLaughlin, S. D. Ordille, M. S. Beltz, M. V. Jarowenko, and I. S. Zagon, "Human Renal Cell Cancer Proliferation in Tissue Culture is Tonically Inhibited by Opioid Growth Factor," *Journal of Urology* 162, no. 6 (December 1999), 2186–91.

33. Personal correspondence between Ian Zagon and Winton Shaer, November 20, 2007, January 10, 2008, and February 12, 2008; personal correspondence between Elaine Moore and Winton Shaer, March 20, 2008.

34. "Opioid Growth Factor (OGF) in Cancer Therapy, A Unique Biotherapeutic Agent," MedInsight Research Institute, October 2006, (http://www.medvision.com/mihpf/medinsight%20-%20ogf%20review.pdf, accessed May 2008).

35. "LDN and Cancer," Lowdosenaltrexone.org, http://www.lowdosenaltrexone.org/ldn_and_cancer.htm (accessed May 18, 2008).

36. "Best-Case Series for the Use of Immuno-Augmentation Therapy and Naltrexone for the Treatment of Cancer," Agency for Healthcare Research and Quality, Evidence Report/Technology Assessment Number 78, April 2003, http://www.ahrq.gov/clinic/epcsums/immaugsum.htm (accessed May 14, 2008).

37. Leslie Costello and Renty Franklin, "Tumor Cell Metabolism: The Marriage of Molecular Genetics and Proteomics with Cellular Intermediary Metabolism; Proceed with Caution!" Commentary, *Molecular Cancer* 5

no. 29 (November 2006), http://www.molec-ular-cancer.com/content/5/1/59 (accessed May 18, 2008).
  38. Authors' correspondence with Ian Zagon, February 16, 2008.

Chapter 6

  1. "Low Dose Naltrexone for Cancers? Why We Are Skeptical," Lymphomation.org, http://www.lymphomation.org/CAM-M-R.htm#LDN (May 18, 2008).
  2. "Autism Fact Sheet," National Institute of Neurological Disorders and Stroke, http://www.ninds.nih.gov/disorders/autism/d etail_autism.htm (accessed May 18, 2008).
  3. "Interview with Professor Jaak Panksepp," Autism Research Institute, March 11, 1997, http://www.autism.org/interview/pank sepp.html (accessed May 18, 2008).
  4. M. Leboyer, M. P. Bouvard, J. Launay, C. Rescasens, M. Plumet, and D. Waller-Per-otte, "Opiate Hypothesis in Infantile Autism? Therapeutic Trials with Naltrexone," En-cephale 19 no. 2 (March–April 1993), 95–102.
  5. K. L. Reichelt, A.-M. Knivsberg, G. Lind, M. Nødland, "Probable Etiology and Possible Treatment of Childhood Autism," Brain Dysfunction 4 (1991); 308–19.
  6. M. Leboyer, et al., "Opiate Hypothesis in Infantile Autism? " 95–102.
  7. M.P. Bouvard, M. Leboyer, J Launay, C. Rescasens, M. Plumet, D. Waller-Perotte, F. Tabuteau, et al., "Low-dose Naltrexone Effects on Plasma Chemistries and Clinical Symptoms in Autism: A Double-blind, Placebo-controlled Study," Psychiatry Research 58 no. 3 (October 1995), 191–201.
  8. M. Campbell, "Naltrexone in Autistic Children: Behavioral Symptoms and Attentional Learning," Journal of the American Academy of Child and Adolescent Psychiatry 32, no. 6 (1993), 1283–91.
  9. Jaquelyn McCandless, Children with Starving Brains: A Medical Treatment Guide for Autism Spectrum Disorders, 3rd ed. (Viroqua, WI: Bramble Books, 2007).
  10. A. Vojdani, M. Bazargan, E. Vojdani, J. Samadi, A Nourian, N. Eghbalieh, and E. Cooper, "Heat Shock Protein and Gliadin Peptide Promote Development of Peptidase Antibodies in Children with Autism and Patients with Autoimmune Disease," Clinical Diagnostic Laboratory Immunology 11, no. 3 (May 2004), 515–24.
  11. C. Molloy, A. Morrow, J. Meinzen-Derr, K. Schleier, K. Dienger, P. Manning-Courtney, M. Altaye, and M. Wills-Karp, "Elevated Cytokine Levels in Children with

Autism Spectrum Disorder," Journal of Neu-roimmjunology 172 (2006), 198–205.
  12. H. Jyonouchi, L. Geng, A. Ruby, and B. Zimmerman-Bier, "Dysregulated Innate Immune Responses in Young Children with Autism Spectrum Disorders: Their Relationship to Gastrointestinal Symptoms and Dietary Intervention," Neuropsychobiology 51, no. 2 (2005), 77–85.
  13. "Growth Regulation and Related Applications of Opioid Antagonists, U.S. Patent 5266574," Ian S. Zagon and Patricia J. McLaughlin, inventors, Wiki Patents Community Patent Review, November 30, 1993, http://www.wikipatents.com/5266574.html (accessed May 20, 2008)
  14. G. M. Eichaar, N. M. Maisch, L. M. Augusto, H. J. Wehring, "Efficacy and Safety of Naltrexone Use in Pediatric Patients with Autistic Disorder," Annals of Pharmacotherapy 40. no. 6 (June 2006), 1086–1095.
  15. J. P. Reneric and M. P. Bouvard, "Opioid Receptor Antagonists in Psychiatry: Beyond Drug Addiction," CNS Drugs, 10 no. 5 (November 1998), 365–82.
  16. Annabel McQuillain, "Why Do Patients Meeting Criteria for Borderline Personality Disorder Deliberately Harm Themselves? Some Hypothesized Neurobiological Correlates," Schweiz Archives of Neurology and Psychiatry, 155 (2004), 212–6.

Chapter 7

  1. Ian Zagon, correspondence with authors, February 16, 2008.
  2. "Growth Regulation and Related Applications of Opioid Antagonists," Wiki Patents Community Patent Review.
  3. Ian Zagon, Joseph Sassani, and Patricia McLaughlin, "Insulin Treatment Ameliorates Impaired Corneal Reepitheliazation in Diabetic Rats," Diabetes 55 (April 2006), 1141–47.
  4. I. S. Zagon, J. Jenkins, J. Sassani, J. Wylie, T. Ruth, J. Fry, C. Lang, and P. Mc-Laughlin, "Naltrexone, an Opioid Antagonist, Facilitates Reepitheliazation of the Cornea in Diabetic Rat," Diabetes 51 (October 2002), 3055–62.
  5. Ian Zagon, Matthew Klocek, Joseph Sassani, David Mauger, and Patricia McLaughlin, "Corneal Safety of Topically Applied Naltrexone," Journal of Ocular Pharmacology and Therapeutics 22, no. 5 (2006), 377–387.
  6. "Method of Treating Chronic Herpes Virus Infections Using an Opiate Receptor Antagonist," Bernard Bihari and Finvola Drury, inventors, October 18, 1994, Freepatentsonline, http://www.freepatentsonline.com/5356900.html (accessed May 15, 2008).

Notes—Chapter 8 165

7. Nicholas Plotnikoff, "Overview of Metenkephalin Actions," NCI Conference, "Low Dose Opioid Blockers, Endorphins and Metenkephalins: "Promising Compounds for Unmet Medical Needs," Bethesda, Maryland, April 20, 2007.

8. R. Bowden, S. Tate, S. Soto, and S. Specter, "Alterations in Cytokine Levels in Murine Retrovirus Infection: Modulation by Combination Therapy, *International Journal of Immunopharmacology* 21, no.12 (December 1999), 815–27.

9. Nicholas Plotnikoff, "Methionine Enkephalin, A New Cytokine with Antiviral and Anti-tumor Properties," chapter 19 in *Potentiating Health and the Crisis of the Immune System* e, ed. A. Mizrahi, Stephen Fulder, and Nimrod Sheinman (New York: Springer, 1997), 193–194.

10. *LDN: The Latest News*, http://www.lowdosenaltrexone.org/ldn_latest_news.htm (accessed May 14, 2008).

11. Kenneth Singleton, *The Lyme Disease Solution* (South Lake Tahoe, CA: BioMed Publishing Group, 2008); personal correspondence with authors, February 22, 2008.

12. I. S. Zagon, and P. J. McLaughlin, "An Opioid Growth Factor Regulates the Replication of Microorganisms," *Life Sciences* 50, no. 16 (1992), 1179–1187.

13. "Growth Regulation and Related Applications of Opioid Antagonists," Wiki Patents Community Patent Review.

14. P. L. Bigliardi, H. Stammer, G. Jost, T. Rufli, S. Buchner, and M. Bigliardi-Oi, "Treatment of Pruritis with Topically Applied Opiate Receptor Antagonist," *Journal of the American Academy of Dermatology* 56, no. 6 (June 2007), 979–88.

15. E. A. Jones, J. Neuberger, and N.V. Bergasa, Opiate Atagonist Therapy for the Pruritus of Cholestasis: The Avoidance of Opioid Withdrawal-like Reactions," *QJM* 95 (2002), 547–52.

16. Ian Zagon, correspondence with Elaine Moore, March 22, 2008.

Chapter 8

1. David Gluck, Introductory Overview, NCI Conference: "Low Dose Opioid Blockers, Endorphins and Metenkephalins "Promising Compounds for Unmet Medical Needs," Bethesda, Maryland, April 20, 2007.

2. Rafael Franco, Rodrigo Pacheco, Carmen Liuis, Gerard Ahern, and Petra O' Connell, "The Emergence of Neurotransmitters as Immune Modulators," *Trends in Immunology* 28, no. 9 (September 2007). 400–407.

3. Yehuda Shavit, Antoine Depaulis, Fredricka Martin, Gregory Terman, Robert Pechnick, Cynthia Zane, Robert Gale, and John Liebeskind, "Involvement of Brain Opiate Receptors in the Immune-suppressive Effect of Morphine," *Proceedings of the National Academy of Science* 83 (September 1986), 7114–7117.

4. Ian Zagon and Patricia McLaughlin, "Opioid Growth Factor Receptor is Unaltered with the Progression of Human Pancreatic and Colon Cancers," *International Journal of Oncology* 29, no. 2 (August 2006), 489–494.

5. J. Wybran, T. Appelboom, J. P. Famaey, and A. Govaerts, "Suggestive Evidence for Receptors for Morphine and Methionine-enkephalin on Normal Human Blood T Lymphocytes, " *Journal of Immunology* 123, no.3 (September 1979), 1068–70.

6. Louis Sanford Goodman, Alfred Gilman, Laurence L. Brunton, John S Lazo, and Keith L. Parker, eds., *Goodman & Gilman's The Pharmacological Basis of Therapeutics*, 11th edition, ed. Laurence L. Brunton (New York: McGraw-Hill, 2006), 550–555.

7. Samira Kiani, Mohammad Ebrahimkhani, Ahmad Shariftabrizi, Behzad Doratotaj, et al., "Opioid System Blockade Decreases Collagenase Activity and Improves Liver Injury in a Rat Model of Cholestasis," *Journal of Gastroenterology and Hepatology* 22, no.3 (March 2007), 406–413.

8. Jorge Daruna, *Introduction to Psychoneuroimmunology* (Burlington, MA: Elsevier Academic Press, 2004), 59–60.

9. Ibid., 69.

10. "Dr. Kamau B. Kokayi Interviews Dr. Bihari, September 23, 2003, WBAI in New York City, Global Medicine Review," http://www.gazorpa.com/interview.html (accessed July 10, 2007).

11. Bernard Bihari, Finvola Drury, Vincent Ragone, Gennaro Ottomanelli, Elena Buimovici-Leine, Milagros Orbe, William Foeste, Johnny Thomas, and Robert Kirt, "Low Dose Naltrexone in the Treatment of Acquired Immune Deficiency Syndrome," Oral Presentation at the IV International AIDS Conference in Stockholm, June 1988, lowdosenaltexone. org, http://www.lowdosenaltrexone.org/ldn_aids_1988.htm (accessed May 20, 2008).

12. Ibid.

13. "LDN and HIV/AIDS," lowdosenaltrexone.org, http://www.lowdosenaltrexone.org/ldn_and_hiv.htm#background (accessed November 20, 2007).

14. G. Gekker, J. Lokensgard, P. Peterson, "Naltrexone Potentiates Anti-HIV-2 Activity of Antiretroviral Drugs in CD4+ Lymphocyte Cultures, *Drug and Alcohol Dependence* 64, no.3 (November 2001), 257–263.

15. "Low-dose Naltrexone and Lipodystrophy," The Body, Seattle Treatment Ed-

ucation Project, May 19, 2000, http://www.the-body.com/content/art2019.html (accessed October 15, 2007).

16. "Naltrexone—A Unique Immune Regulator Showing Promise in All Manner of Autoimmune Diseases Including HIV," International Aging Systems, http://www.antiagingsystems.com/a2z//naltrexone.htm (accessed July 29, 2007).

17. Nicholas Plotnikoff, "Overview of Metenkephalin Actions," NCI Conference, "Low Dose Opioid Blockers, Endorphins and Metenkephalins: Promising Compounds for Unmet Medical Needs," Bethesda, Maryland, April 20, 2007."

Chapter 10

1. M. Gironi, R. Furlan, M Rovaris, G Comi, M. Fillipi, A. Panerai, and P. Sacerdote, "Beta Endorphin Concentrations in PBMC of Patients with Different Clinical Phenotypes of Multiple Sclerosis," *Journal of Neurology and Neurosurgical Psychiatry* 74 (2003), 495–7.

2. K. J. Smith, R. Kapoor, P. A. Felts, "Demyelination: The Role of Reactive Oxygen and Nitrogen Species," *Brain Pathology* 9, no.1 (January 1999), 69–92.

3. Jorge H. Daruna, *Introduction to Psychoneuroimmunology* (London: Elsevier Academic Press, 2004), 69.

4. Daniel Klein and Alexander Tabarrok, "Do Off Label Drug Practices Argue Against FDA Efficacy Requirements? Testing an Argument by Structured Conversations with Experts," http://mason.gmu.edu/~atabarro/Do%20Off-Label%2024.pdf (accessed May 20, 2008).

5. Michael Doonan, "The Economics of Prescription Drug Pricing," Background Paper, Council on the Economic Impact of Health System Chance, National Press Club, Washington, DC, March 28, 2001.

6. Joe Undergrad, "Coping with an Unpopular Cure," *Columbia Spectator*, May 1, 2004.

7. State of Vermont, Joint House Resolution, JRH-6, 2005–2006, www.leg.state.vt.us/docs/legdoc.cfm?URL=/docs/2006/resolutn/JRH006.htm (accessed May 20, 2008).

8. Zorica Mancev, Gordana Pesic, Stanislava Stanojevic, and Jelena Radulovic, "The Immunomodulating Effects of Specific Opioid Antagonists after Their Intracerebroventicular Application," *University Medicine and Biology* 7, no.1 (2000), 26–30.

9. "MS Hope from Heroin Addiction Drug," in "Low Dose Naltrexone—The Latest News," LDN Research Trust, http://www.msrc.co.uk/index.cfm/fuseaction/show/page id/1306 (accessed May 20, 2008).

10. Nikola Stambuk, "Cytogenetic Effects of Met-enkepahlin (Peptid-M) on Human Lymphocytes," *Croatica Chemica Acta* 71 (1998), 591–605, précis: http://public.carnet.hr/ccacaa/591.html (accessed May 20, 2008).

11. S. A. Eisen, et al., "Gulf War Veterans' Health: Medical Evaluation of a U.S. Cohort," *Annals Internal Medical* 142 (2005), 881–90.

12. W. K. Hallman, et al., "Symptom Patterns among Gulf War Registry Veterans," *American Journal of Public Health* 93 (2003), 624–30.

13. H. V. Thomas, et al,. "Pain in Veterans of the Gulf War of 1991: A Systematic Review," *BMC Musculoskeletal Disorders* 7 (2006), 74.

14. H. V. Thomas, et al., "Systematic Review of Multi-symptom Conditions in Gulf War Veterans," *Psychological Medicine* 36 (2006), 735–47.

15. M. B. Yunus, "Central Sensitivity Syndromes: A New Paradigm and Group Nosology for Fibromyalgia and Overlapping Conditions, and the Related Issue of Disease versus Illness," *Seminars in Arthritis and Rheumatism* 2008, "epub ahead of print," PubMed, http://www.ncbi.nlm.nih.gov/pubmed/18191990 (accessed May 20, 2008).

16. Jarred Younger, correspondence with author Elaine Moore, March 27, 2008.

Glossary

ACTH: *See* **Adrenocorticotropic hormone.**

Acetylcholine: A neurotransmitter that appears to be involved in learning and memory, severely diminished in the brains of persons with Alzheimer's disease.

Acquired Immunodeficiency Syndrome (AIDS): A life-threatening disease caused by the human immunodeficiency virus (HIV) and characterized by breakdown of the body's immune defenses.

Adrenal glands: Glands located above each kidney that secrete hormones that help the body to withstand stress and regulate metabolism.

Adrennocorticotropic hormone (ACTH): Hormone secreted by the anterior pituitary gland that stimulates the adrenal glands to produce cortisone.

Affinity: The strength of the interaction between a ligand and a receptor. When two ligands exist at equal concentration, the ligand whose affinity is higher will tend to displace the other from a receptor, assuming the low-affinity ligand is bound reversibly to the receptor.

Agonist: A molecule that activates a cell receptor and triggers a biological response.

AIDS: *See* **Acquired Immunodeficiency Syndrome.**

Alzheimer's disease: A chronic, progressive, neurodegenerative disorder characterized by impairment of memory, thinking, behavior, and emotion; the most common cause of dementia and the fourth-leading cause of death in the United States.

Amyloid beta protein: A generic name for a class of sticky proteins found in the brains of patients with Alzheimer's disease; also known as beta-amyloid, A beta, or beta-amyloid protein.

Anergy: A state of unresponsiveness, induced when the T cell's antigen receptor is stimulated, that effectively freezes T cell responses pending a "second signal" from the antigen-presenting cell (co-stimulation).

Angiogenesis: The formation of new blood vessels. Angiogenesis is essential for the growth of tumors. Tumor cells release chemicals to encourage blood vessel growth.

Animal model: A laboratory animal useful for medical research because it has specific characteristics that resemble a human disease or disorder. Scientists can create animal models, usually laboratory mice, by transferring new genes into them.

Antagonist: A molecule that blocks the ability of a given chemical to bind to its receptor, preventing a biological response.

Antibody: A protein produced and secreted by B cells in response to a foreign antigen, as part of the body's defense against disease; it neutralizes the antigen by binding to it.

167

Antibody-dependent cell-mediated cytotoxicity (ADCC): An immune response in which antibodies, by coating target cells, make them vulnerable to attack by immune cells.

Antigen: Any substance that, when introduced into the body, is recognized by the immune system. Most foreign (to the body) antigens are proteins such as pollen or infectious agents.

Antigen-presenting cells: B cells, cells of the monocyte lineage (including macrophages as well as dendritic cells), and various other body cells that "present" antigen in a form that T cells can recognize.

Aphasia: A condition of inability to use language caused by destruction of the brain cells that govern language.

Apoptosis: Programmed cell death ("suicide"); different types of cells are programmed to die at specific times in a process of apoptosis.

Arachidonic acid: A fatty acid that helps make up cell walls, which is metabolized to eicosanoids, a family of substances highly active in immune and inflammatory responses.

Astrocytes: Large star-shaped glial cells found in the brain, which regulate the composition of the fluids that surround nerve cells; these can also form scar tissue after injury.

Ataxia: The lack of coordination and unsteadiness that result from the brain's failure to regulate the body's posture and the strength and direction of limb movements. Ataxia is most often caused by disease activity in the cerebellum.

Autoimmune disease: A disease in which the body produces an immunogenic (immune system) response to some constituent of its own tissue; the immune system loses its ability to recognize some tissue or system within the body as "self" and targets and attacks it as if it were foreign.

Autonomic: The part of the nervous system which controls the tissues, organs and systems without conscious thought; concerned with reflex control of bodily functions (for example, heartbeat and blood pressure); a functional division based on the role that autonomic neural pathways play, regardless of whether these cross through the CNS or the PNS (the two divisions: sympathetic and parasympathetic).

B cells: Small white blood cells crucial to the immune defenses. Also known as B lymphocytes, they are derived from bone marrow and develop into plasma cells that are the source of antibodies.

Beta-endorphins: Neuropeptides that bind to opioid receptors (as an agonist) in the brain and have potent analgesic (painkiller) ability.

Biological response modifiers: Substances, either natural or synthesized, that boost, direct, or restore normal immune defenses. BRMs include interferons, interleukins, thymus hormones, and monoclonal antibodies.

Blastogenesis: The transformation of small lymphocytes into larger cells capable of undergoing mitosis.

Blood-brain barrier: Protects the brain from "foreign substances" in the blood that may injure the brain, protects from hormones and neurotransmitters in the rest of the body, and maintains a constant environment for the brain. It is semi-permeable; that is, it allows some materials to cross, but prevents others from crossing.

Bone marrow: Soft tissue located in the cavities of the bones. The bone marrow is the source of all blood cells.

CD: *See* **Cluster Designation.**

CD4 T Cells: T cells with molecules called CD4 on its surface. These "helper" cells initiate the body's response to invading microorganisms such as viruses.

CD8 T Cells: Suppressor T cells with molecules called CD8 on their surface. CD8 cells destroy cells that have been infected with foreign invading microorganisms. CD8 cells also produce antiviral substances (antibodies) that help fight off foreign invaders.

CNS: *See* **Central nervous system**

Caspase: any of a group of proteases that mediate apoptosis.

Cells: The basic building blocks of human tissue that can perform all of the functions of the corresponding tissue, such as signaling and growth.

Cellular immunity: Immune protection provided by the direct action of immune cells (as distinct from soluble molecules such as antibodies).

Central nervous system: Represents the largest part of the nervous system, including the brain and the spinal cord.

Chemokine: Soluble molecules that chemically attract lymphocytes and other cells.

Cholestasis: Impairment of bile flow due to obstruction in small bile ducts (intrahepatic cholestasis) or obstruction in large bile ducts (extrahepatic cholestasis).

Chromosomes: Physical structures in the cell's nucleus that house the genes. Each human cell has twenty-three pairs of chromosomes.

Chronic inflammation: Overproduction of free radicals by the immune system results in inflammatory-related disease such as arthritis, arteriosclerosis, heart attack, Type II diabetes, lupus, MS, asthma, and inflammatory bowel disease.

Clone: A genetically homogeneous population derived from a single cell

Cluster Designation: International nomenclature for cell surface molecules (CD number).

Collagen: The major structural fibrous protein that gives the skin and other connective tissue strength and resilience.

Complement: A complex series of blood proteins whose action complements the work of antibodies. Complement destroys bacteria, produces inflammation, and regulates immune reactions.

Cortisone: A glucocorticoid steroid hormone, produced by the adrenal glands or synthetically, that has anti-inflammatory and immune-system suppressing properties. Prednisone and prednisolone also belong to this group of substances.

Cytokine: Secreted signaling molecules that allow communication between lymphocytes and other cells. Cytokines include lymphokines, produced by lymphocytes, and monokines, produced by monocytes and macrophages.

Cytolysis: The dissolution or destruction of a cell.

Delta receptors: The d-Opioid receptors are a class of opioid receptors with enkephalins as their endogenous ligands, and which possess features not present in the mu receptors or kappa receptors. Activation of delta receptors produces some analgesia, although less than that of mu-opioid agonists, and antidepressant effects.

Demyelinating disease: A disease causing loss or damage to the protective myelin sheath covering nerve fibers; symptoms are determined by the functions normally contributed by the affected neurons.

Dendrites: Branched extensions of the nerve cell body that receive signals from other nerve cells. Each nerve cell usually has many dendrites.

Dendritic cell: Immune cells that process antigen material and present it to other cells of the immune system, thus functioning as antigen-presenting cells. They typically use threadlike tentacles to enmesh antigens.

Dopamine: A precursor of epinephrine and norepinephrine, it is a major neurotransmitter present in regions of the brain that regulate movement, emotion, motivation, and the feeling of pleasure.

Dopaminergic: Meaning "related to the neurotransmitter dopamine." A synapse is dopaminergic if it uses dopamine as its neurotransmitter. A substance is dopaminergic if it is capable of producing, altering, or releasing dopamine.

Dynorphins: A class of opioid peptides, which prefer kappa-opioid receptors and have been shown to play a role as central nervous system transmitters. Other opioid peptides include beta-endorphin, and met-enkephalin.

Dystonia: A neurological movement disorder in which sustained muscle contractions cause twisting or repetitive movements or abnormal posture.

EPA DHA: Essential fatty acids (eicosapentaenoic acid and docosahexaenoic acid), long chain polyunsaturated omega-3 fats found in coldwater fish, commonly used as a supplement. Essential for brain growth and function, they are converted into hormone-like substances called prostaglandins, which regulate cell activity.

Eicosanoids: A family of powerful, hormone-like compounds derived from oxygenation of essential fatty acids (EFAs). These compounds include prostaglandins, leukotrienes, and thromboxanes, which are responsible for many of the beneficial effects of essential fatty acid oils.

Effector cell: Cell that has developed full immune functions (for example, a cytotoxic /killer cell).

Endocrine System: A system of ductless glands that regulate bodily functions via hormones secreted into the bloodstream.

Endogenous: Produced by the body itself.

Endogenous opiates: An internally produced (within the brain or other body tissues) morphine-like substance that results in a feeling of euphoria. Includes endorphins, enkephalins, and dynorphins.

Endorphins: Natural endogenous opioids manufactured in the brain that reduce sensitivity to pain and stress.

Enkephalins: Endogenous ligands, or specifically opioids, that are internally derived and bind to the body's delta-opioid receptors to regulate pain. Both are products of the proenkephalin gene.

Enzymes: Protein made by the body that brings about a chemical reaction; for example, the enzymes produced by the gut to aid digestion.

Epidemiology: The study of factors affecting the health and illness of populations, serves as the foundation and logic of interventions made in the interest of public health and preventive medicine.

Essential Fatty Acids (EFA): Long-chain, polyunsaturated acids (linoleic, linolenic, and arachidonic) of the omega-6 and omega-3 families, which are essential for maintaining health; they cannot be synthesized by the body but must be ingested. EFAs are a major component of cellular membranes and have vital functions in nearly every metabolic function in the body.

Eukaryote: A cell with a membrane-bound nucleus. The vast majority of species (plants, animals, protozoa) are eukaryotic. *See also* **Prokaryote.**

Exacerbation: A worsening of symptoms or a disease progression; attack or flare of symptoms.

Excitotoxins: Brain chemicals (neurotransmitters) that stimulate neurons to fire; in excessive amounts, excitotoxins can damage neurons.

Exogenous: Originating or produced outside of the body.

Fibrosis: The growth of fibrous scar tissue, possibly due to infection, inflammation, injury, or even healing.

FoxP3: A master gene for controlling regulatory T cell functions.

Glial cells: Including microglia and astrocytes, these are the resident immune system cells of the central nervous system.

Glutamate: An amino acid neurotransmitter normally involved in learning and memory. In excess it can be an excitotoxin and appears to cause nerve cell death in a variety of neurodegenerative disorders.

Growth hormone: A polypeptide hormone secreted by the anterior lobe of the pituitary gland that promotes growth of the body.

HIV: *See* **Human Immunodeficiency Virus.**

Histamine: A depressor amine found in plant and animal tissue, derived from histidine and released from mast cells in the immune system as part of an allergic reaction. It is a powerful stimulant of gastric secretion, constrictor of bronchial smooth muscle, and vasodilator.

Homeostasis: The ability or tendency of an organism or cell to maintain internal equilibrium by adjusting its physiological processes.

Hormone: A chemical or protein, secreted by the glands of the endocrine system and released into the bloodstream, that regulates bodily functions by acting as a messenger or stimulatory signal, relaying instructions to stop or start certain physiological processes.

Human Immunodeficiency Virus (HIV): A retrovirus that causes AIDS by infecting helper T cells of the immune system.

IL-10: Interleukin-10; a cytokine that downregulates antiviral responses.

IL-2: Interleukin-2; a lymphokine required by activated T cells for growth.

IL-2R: Interleukin-2 receptor expressed on activated T and B lymphocytes.

IL-4: Interleukin-4; a cytokine produced by T cells to help antibody responses.

Idiotype: The unique appearance of an antibody binding site to the immune system.

Immunoglobulin: A general term for antibodies made by B cells, which bind onto invading organisms, leading to their destruction.

In Vitro: Literally "in glass"; a study conducted in a vessel (glass container or diffusion cell) using excised tissues, enzyme preparations or microorganisms.

In Vivo: Literally "in life"; a study conducted using living, intact organisms.

Inflammatory response: A normal process (as opposed to chronic inflammation), caused by a burst of free radicals produced by the immune system in response to an injury. Functions to prevent infection and to promote healing at a site of injury.

Interferon: A biological response modifier cytokine that interferes with the division of cancer cells with the potential to slow tumor growth. There are several types of interferons, including interferon-alpha, -beta, and -gamma

Interleukin: Interleukins are a group of cytokines that were first seen to be expressed by white blood cells (leukocytes, hence the -*leukin*) as a means of communication (*inter*).

Kappa receptors: The κ-opioid receptors are a class of opioid receptors with dynorphins as the primary endogenous ligands, which possess features not present in the mu receptors or delta receptors. Functions include mediation of the hallucinogenic side effects of opioids, sedation and miosis.

Leukocyte: Any of the various white blood cells that together make up the immune system. Neutrophils, lymphocytes, and monocytes are all subtypes of leukocytes.

Leukotrienes: Naturally produced eicosanoid lipid mediators, released by mast cells during an allergic response or asthma attack, and usually involving the production of histamine. Produced in the body from arachidonic acid.

Ligand: A molecule which binds specifically to a receptor to form a complex (enkephalins are the ligands that bind to delta-opioid receptors).

Lipodystrophy: A disturbance of fat metabolism that involves the absence of fat and/or the abnormal distribution of fat in the body. The term is commonly used to refer to any type of body fat redistribution, including accumulations of visceral, abdominal, and breast adipose tissue.

Low Dose Naltrexone (LDN): Reduced dose of naltrexone in a range around 4.5mg, which exerts its primary effects by increasing production of the endorphin met-5-enkephalin.

Lymph node: A lymph organ draining tissues distributed widely throughout the body and linked by lymphatic vessels. This is where immune responses are initiated, including those of B and T Lymphocytes and other immune cells.

Lymphokine: Powerful chemical substances secreted by lymphocytes. These soluble molecules help direct and regulate the immune responses.

Lymphoma: Any of various usually malignant tumors, such as Hodgkin's lymphoma, that arise in the lymph nodes or other lymphatic tissues, often manifested by painless enlargement of one or more lymph nodes.

MBP: *See* **Myelin basic protein.**

Macrophage: A large and versatile immune cell that acts as a microbe-devouring phagocyte, an antigen-presenting cell, and an important source of immune secretions.

Mast cells: Cells which synthesize and store histamines, are found in most body tissues, and play an important role in the body's allergic response.

Melanocyte stimulating hormone (MSH): A class of peptide hormones produced by cells in the intermediate lobe of the pituitary gland. MSH stimulates the production and release of melanin (melanogenesis) by melanocytes in skin and hair. Also produced by a subpopulation of neurons in the arcuate nucleus of the hypothalamus; when released into the brain by these neurons MSH affects appetite and sexual arousal.

Memory: The basis of adaptive immunity such that T or B lymphocyte responses to previously encountered antigens are more vigorous on subsequent exposures.

Met-enkephalin: An endorphin with potent antiviral and antitumor properties due to its ability to increase production of interleukin-2, NK lymphocytes, CD4 lymphocytes, and CD 8 lymphocytes. Produced from its own precursor, proenkephalin.

Methadone: A synthetic opioid, used medically as an analgesic, antitussive and a maintenance anti-addictive for use in patients on opioids (it prevents withdrawal syndrome and blocks the "high"). Although chemically unlike morphine or heroin, methadone also acts on the opioid receptors and thus produces many of the same effects.

Microglia: Specialized immune cells, related to macrophages, that protect the central nervous system.

Mitogen: A chemical, usually some form of a protein, that encourages a cell to commence cell division, triggering mitosis.

Monoclonal antibody: An antibody with a single specificity produced by an immortal B cell line.

Monocyte: A large white blood cell that can leave the blood vessels and enter body tissue, at which point it becomes a macrophage.

Motor fibers: Motor fibers are efferent nerve fibers that transmit impulses from the central nervous system to muscles, causing them to contract; produced in the anterior gray horn of the spinal cord.

Mu receptors: A class of opioid receptors with high affinity for enkephalins and beta-endorphins but low affinity for dynorphins. The opiate alkaloids morphine and codeine are known to bind to this receptor, causing euphoria or a drug "high." Respiratory depression is the primary way that overdoses kill; opioid overdoses can be rapidly reversed with opioid antagonists (such as naloxone or naltrexone).

Myelin basic protein (MBP): The primary constituent of the protective myelin sheath that covers nerve axons; MBP is also expressed within the nervous system in other locations, and it may be seen in lymph and thymus tissue.

Myelinated nerve fibers: Nerve fibers surrounded by myelin sheaths; the myelin is produced by supporting cells known as oligodendrocytes within the central nervous system and Schwann cells in the peripheral nervous system.

Naive: The state of lymphocytes before first exposure to their specific antigen

Naloxone: An opioid antagonist, similar to naltrexone, but with poor bio-availability when taken sublingually. Brand name: Narcan.

Naltrexone: An opioid receptor antagonist used primarily in the management of alcohol dependence and opioid dependence. It is marketed in generic form as its hydrochloride salt, naltrexone hydrochloride (NTX), and is also marketed using the trade name Revia as a 50-mg tablet.

Natural killer (NK) cell: A T lymphocyte that turns off specific immune responses

Narcotic: literally "sleep- or stupor-inducing agent"; a term applied indiscriminately to describe any exogenous compound with a sedating profile. Use of the term with reference to opiates is not recommended, due to its ambiguity.

Neurites: Nerve fiber processes, including both axons and dendrites, that extend from the cell body of neurons; responsible for both receiving and transmitting information.

Neutrophils: The predominant type of white blood cell.

Neuroblastoma: A malignant tumor composed of neuroblasts (cells from which nerve cells develop), originating in the autonomic nervous system or the adrenal medulla and occurring chiefly in infants and young children.

Neurodegeneration: Progressive loss of structure or function of neurons, including death of neurons.

Neuroendocrine: Having to do with the interactions between the nervous system and the endocrine system. *Neuroendocrine* describes certain cells that release hormones into the blood in response to stimulation of the nervous system.

Neurogastroenterology: A research area in the field of gastroenterology studying interactions of the central nervous system (brain) and the gut—the so-called brain-gut axis. Research focuses upon upward (sensory) and downward (motor and regulatory) neural connections, and endocrine influences on gut function. Another area of study is the enteric nervous system in itself. Clinical research deals mainly with motility disorders and functional bowel disorders (for example, irritable bowel syndrome).

Neuromodulator: As opposed to direct synaptic transmission in which one neuron directly influences another neuron, neuromodulatory transmitters secreted by a small group of neurons diffuse through large areas of the nervous system, having an effect on multiple neurons. Examples of neuromodulators include dopamine, serotonin, acetylcholine, and histamine.

Neurons: Cells of the nervous system that consist of a cell body, the soma, that contains the nucleus and the surrounding cytoplasm, several short, thread-like projections (dendrites), and one long filament (the axon).

Neuropeptide: Any of the variety of peptides found in neural tissue (for example, nerurotransmitters, endorphins, enkephalins), that function as neuromodulators in the nervous system and as hormones in the endocrine system.

Neurotransmitter: Any endogenous compound that plays a role in synaptic nervous transmission.

Neurotrophic: Having a selective affinity for nerve tissue.

Nervous system: Gross anatomy components (parts that are large enough to be seen with the naked eye) and microanatomy (the system at a cellular level) that are further divided into two parts: the central nervous system (CNS) and the peripheral nervous system (PNS).

Nitric Oxide: A free-radical gas produced endogenously by a variety of cells. It is synthesized from arginine by a complex reaction, catalyzed by nitric oxide synthase (NOS). Nitric oxide is an endothelium-derived relaxing factor.

OGF-OGFr complex: Repairs tissue and helps the body heal itself or return to a homeostatic state; low dose naltrexone increases the production of this complex.

Oligodendrocytes: A neuronal support or glial cell that produces insulating myelin.

Oligopeptide: Consists of a small number (*oligo-*, "few") of amino acids linked together, as opposed to a polypeptide (*poly-*, "many").

Omega 3: The name given to a family of polyunsaturated essential fatty acids of which DHA and EPA are members, found in high amounts in coldwater fish. Recommended to be consumed in balance with Omega 6, at a ratio of 1–2:1 (omega 3:omega 6).

Omega 6: The name given to a family of polyunsaturated essential fatty acids, derived from linoleic acid (LA). Abundant in vegetable oils such as safflower and sunflower oils.

Opiates: A major class of highly addictive drugs that depress the central nervous system and are used medically to relieve pain, but are frequently abused and trafficked illegally. Examples include morphine, heroin, fentanyl, methadone and codeine.

Opioid: An endogenous peptide or exogenous compound (drug), which possesses some affinity for any or all of the opioid receptor subtypes (delta, kappa, mu).

Opioid growth factor (OGF): A name designated for the endogenous pentapeptide methionine enkephalin, [Met5]-enkephalin, by Ian Zagon of Penn State University, when his team discovered its function as a growth factor.

Opioid growth factor receptor (OGFr): Identified by Ian Zagon as the receptor activated by OGF.

Opioid receptor: Molecular structures on the cells that respond to—or are influenced by—opiate and endorphin molecules, and which occur in brain and spinal cord cells, immune system cells, the gastrointestinal tract, as well as in most cancer-tumor cells.

Opioidergic: Having opioid-like properties.

Oxytocin: A short polypeptide hormone, $C_{43}H_{66}N_{12}O_{12}S_2$, released from the posterior lobe of the pituitary gland, that stimulates the contraction of smooth muscle of the uterus during labor and facilitates ejection of milk from the breast during nursing.

Pentapeptide: A polypeptide that contains five amino-acid residues.

Peptides: Short polymers formed from the linking of amino acids. The link between one amino-acid residue and the next is known as an amide bond or a peptide bond. Peptides combine to make proteins, including antigens.

Peripheral nervous system (PNS): Consists of all the other nervous structures that do not lie within the CNS, which serve the limbs and organs; divided into the somatic nervous system and the autonomic nervous system.

Peroxynitrite: A potent oxidant synthesized by the cell during its normal metabolism, formed from the reaction of two free radicals, nitric oxide (NO) and superoxides. These can damage a wide array of molecules in cells.

Phagocyte: An immune system cell that can surround and kill microorganisms and remove dead cells. Phagocytes include macrophages.

Polypeptide: A chain of amino acids connected by peptide linkages. A polypeptide is longer than an oligopeptide, and may constitute an entire protein.

Prion: A protein molecule that lacks nucleic acid (has no DNA or RNA), often considered to be the cause of various infectious diseases of the nervous system, such as Creutzfeldt-Jakob disease or mad cow disease, and scrapie. These are very resilient, not easy to kill.

Prokaryote: An organism (mycoplasma, blue-green alga, bacterium), whose cells contain no membrane-bound nucleus or other membranous organelles. *See also* **Eukaryote.**

Prolactin: A pituitary hormone that stimulates and maintains the secretion of milk.

Proopiomelanocortin (POMC): A hormone complex synthesized by the anterior pituitary gland, hypothalamus, brainstem, and melanocytes in the skin. Metabolized into four separate hormones: adrenocorticotropic hormone (ACTH), melanocyte stimulating hormone (MSH), met-enkephalin and beta-endorphin.

Prostaglandins: Any of a class of hormone-like, fat-soluble, regulatory molecules made from fatty acids such as arachidonic acid; prostaglandins have strong physiological effects, such as the constriction or relaxation of vascular smooth muscle, aggregation of platelets, and causing pain in spinal neurons. Their production is blocked by NSAIDs (non-steroidal anti-inflammatory drugs).

Pruritis: Itching of the skin, sometimes accompanied by a rash, which may be associated with various types of cancers, cancer treatments and other medications.

Psychoneuroimmunology (PNI): The study of the interaction between psychological processes and the nervous and immune systems of the human body. The main interest of PNI is the interaction between the nervous and immune systems, and the relation between mind processes and health.

Reactive oxygen species (ROS): Damaging molecules, including oxygen radicals and other highly reactive forms of oxygen that are potential sources of damage.

Retrovirus: A virus that needs cells from a "host" in order to make more copies of itself (replication). In the case of HIV, CD4 cells are the host cells that aid HIV in replication.

Selectivity: The relationship between the affinity of a compound for a particular receptor and its affinity for other types of opioid receptor. For instance, a compound that will bind with high affinity to the mu-receptors, but with very low affinity to kappa and delta receptors, is said to possess high selectivity for mu.

Serotonin: A neurotransmitter that regulates many functions, including mood, behavior, physical coordination, appetite, body temperature, sleep and sensory perception.

Somatic: The nervous system component responsible for coordinating the body's movements, and also for receiving external stimuli. It is the system that regulates activities that are under conscious control.

Somatostatin: A polypeptide hormone, produced chiefly by the hypothalamus, that inhibits the secretion of various other hormones, such as somatotropin, glucagon, insulin, thyrotropin, and gastrin.

Substantia nigra; Pigmented area of gray matter within the anterior portion of the midbrain that separates the tegmentum from the crus cerebri; the substantia nigra is a large motor nucleus composed of medium size neurons with many processes; these neurons contain inclusion granules of melanin pigment.

Superoxide: A highly toxic free radical that is deployed by the immune system to kill invading microorganisms, and may contribute to the pathogenesis of many diseases. It is continuously removed by the enzyme superoxide dismutase.

Superoxide dismutase (SOD): SOD out-competes damaging reactions of superoxide, thus protecting the cell from superoxide toxicity. It is an important antioxidant defense in nearly all cells exposed to oxygen.

Suppressor T-cell: A T lymphocyte that suppresses (turns off) specific immune responses

Synapse: Where two neurons or a neuron and a skeletal muscle come into close proximity and communicate with one another, communication occurs over a site or bridge known as a synapse. The synaptic cleft refers to the gap or space between neurons' axons and dendrites where functional communication occurs. Synapses may be either chemical or electrical. Synaptic loss occurs in some neurodegenerative diseases and prevents cellular communication.

Synthetic: Produced by a synthesis of elements or materials, especially not of natural origin; human-made.

T-cell receptor: The receptor on a T-cell that binds antigen+MHC and signals recognition.

Th1: T lymphocytes making cytokines to help inflammation and antiviral responses.

Th2: T lymphocytes making cytokines to help antibody responses.

Th3: T lymphocytes making predominantly TGF (and helping IgA antibody responses).

T-lymphocyte: A nucleated white blood cell made in the thymus.

Tr1: T lymphocytes that regulate Th1 responses (may be related to Th3).

Transforming growth factor (TGF): A cytokine that downregulates antigen presentation.

Thromboxanes: Like the leukotrienes, thromboxanes are eicosanoids also formed from arachidonic acid, but using a separate biochemical pathway. Their effects include platelet aggregation, vasoconstriction, and contraction of bronchial smooth muscle.

Thymus: The organ that generates T-lymphocytes, found just above the heart.

Tolerance: A non-aggressive state of the immune system, normally associated with self-recognition.

Treg: Naturally occurring CD4+CD25+ T lymphocytes that regulate responses to self-antigens and inhibit autoimmunity.

Tumor necrosis factor (TNF): A cytokine that induces programmed cell death in cells with a receptor.

Vasopressin: A hormone secreted by the posterior lobe of the pituitary gland that constricts blood vessels, raises blood pressure, and reduces excretion of urine. Also called antidiuretic hormone.

Appendix

Clinical Trials of LDN and Related Compounds

AUTISM

Condition:	Pervasive developmental disorder (PDD)
Study Type:	Human, randomized, double-blind, placebo-controlled, crossover assignment, safety/efficacy study
Status:	Not yet open for participant recruitment, as of March 2008
# Subjects:	Fifty children aged three to six (estimated enrollment)
Medication:	LDN
Where:	Jerusalem Institute for Child Development, Jerusalem, Israel
Researchers:	David S Wilensky
Website:	http://clinicaltrials.gov/ct2/show/NCT00318162
Contact:	David S Wilensky, Phone: 97227828142, e-mail: davidvil@012.net.il
Official Title:	Randomized Double-Blind Trial of Low-Dose Naltrexone for Children With PDD
Primary Outcome Measures:	Play observation, Autism Behavior Checklist (ABC) questionnaire

AUTISM

Study Type:	Human children, thirty-day, double-blind, placebo-controlled
Status:	Completed 1995
# Subjects:	Not specified
Medication:	LDN, oral, 0.5 mg/kg/day
Where:	Service of the Psychopathology of Infants and Adolescents, Robert Debre Hospital, Paris, France
Researchers:	M. P. Bouvard, M. Leboyer, J. M. Launay, et al.
Official Title:	Low-dose Naltrexone Effects on Plasma Chemistries and Clinical Symptoms in Autism: A Double-blind, Placebo-controlled Study [article in French]
Primary Outcome Measures:	Conjoint clinical and biochemical evaluations of therapeutic effects
Published:	Bouvard, M. P., M. Leboyer, J. M. Launay, et al., "Low-dose Naltrexone Effects on Plasma Chemistries and Clinical Symptoms in Autism: A Double-blind, Placebo-controlled Study," *Psychiatry Research* 58 (1995) 191–201

Results   Marginally better overall clinical benefit following naltrexone hy-
                 drochloride, degree of improvement appeared to be related to plasma
                 chemical profiles. The best clinical responders exhibited the clearest
                 normalization of the elevated plasma chemistries, especially in C-ter-
                 minal-beta-endorphin and serotonin
Conclusion: The results suggest that naltrexone hydrochloride only benefits a sub-
                 group of autistic children, who may be identified by the presence of
                 certain plasma abnormalities; these results suggest a possible linkage be-
                 tween abnormal plasma chemistries, especially those related to the pro-
                 opiomelanocortin system, and autistic symptoms

AUTISM

Study Type: Human children, open trial
Status: Completed 1993
Subjects: Two female children
Medication: Naltrexone, oral, 1 mg/kg/day
Where: Service of the Psychopathology of Infants and Adolescents, Robert
                 Debre Hospital, Paris, France
Researchers: M. Leboyer, M. P. Bouvard, J. M. Launay, et al.
Official Title: Opiate Hypothesis in Infantile Autism? Therapeutic Trials with Naltrex-
                 one [article in French]
Rationale: The opioid hypothesis suggests that childhood autism may result from
                 excessive brain opioid activity during neonatal period which may con-
                 stitutionally inhibit social motivation, yielding autistic isolation and
                 aloofness (Panksepp, 1979)
Primary   Clinical evaluations of therapeutic effects
Outcome
Measures:
Published: Leboyer, M., P. Bouvard, J. Launay, C. Rescasens, M. Plumet, D. Waller-
                 Perotte, "Opiate Hypothesis in Infantile Autism? Therapeutic Trials
                 with Naltrexone," Encephale 19 no. 2 (March–April 1993), 95–102
Results   An immediate reduction of hyperactivity, self-injurious behavior and
                 aggressiveness, while attention improved; in addition, social behaviors,
                 smiling, social seeking behaviors and play interactions increased

CANCER

Condition: Cancer
Study Type: Animal studies using chick eggs
Status: Completed
Medication: Opioid growth factor (OGF)
Where: Pennsylvania State University College of Medicine
Researchers: Ian S. Zagon and colleagues
Website: http://fred.psu.edu/ds/retrieve/fred/investigator/isz1
Contact: Phone: 717-531-8650
Published: J. Blebea, J. Mazo, T. Kihari, J. H. Vu, P. J. McLaughlin, R. G. Atnip, I.
                 S. Zagon, "Opioid Growth Factor Modulates Angiogenesis," Journal of
                 Vascular Surgery 32 no. 2 (August 2000), 364–373.
Results: Number of blood vessels and the blood vessel length were decreased in
                 vivo

Conclusion: Opioid growth factor has a significant inhibitory effect on angiogenesis

PANCREATIC CANCER

Study Type: Human, phase 1, to establish the maximum tolerated dose (MTD), and determine safety and toxicity of OGF
Status: Completed
Subjects: Patients with unresectable pancreatic adenocarcinoma
Medication: Opioid growth factor (OGF) treatment, s.c. and i.v. administration
Where: Pennsylvania State University College of Medicine
Researchers: Ian S. Zagon and colleagues
Website: http://fred.psu.edu/ds/retrieve/fred/investigator/isz1
Contact: Phone: 717-531-8650
Published: J. P. Smith, R. I. Conter, S. I. Bingaman, H. A. Harvey, D. T. Mauger, M. Ahmad, L. M. Demers, W. B. Stanley, P. J. McLaughlin, and I. S. Zagon, "Treatment of Advanced Pancreatic Cancer with Opioid Growth Factor: Phase I," *Anticancer Drugs* 15 no. 3 (March 2004), 203–209.
Results: Two subjects had resolution of liver metastases and one showed regression of the pancreatic tumor; mean survival from the time of diagnosis was 8.7 months (range 2–23 months) in the i.v. group and 9.5 months (range 1–18 months) in the s.c. group.
Side Effects: No adverse events were reported
Conclusion: We conclude that OGF can be safely administered to patients with advanced pancreatic cancer; further studies are needed to determine the efficacy of OGF alone or in combination with present modes of therapy for the treatment of pancreatic cancer

PANCREATIC CANCER

Study Type: Human; open-labeled phase 2 interventional treatment
Status: Ongoing; start date October 2003; still enrolling as of February 2008.
Subjects: Fifty
Medication: Opioid growth factor (OGF) treatment, i.v. over forty-five minutes once weekly.
Where: Pennsylvania State University College of Medicine
Researchers: Ian S. Zagon and colleagues
Website: http://fred.psu.edu/ds/retrieve/fred/investigator/isz1
Contact: Phone: 717- 531-8650
Official Title: Treatment of Advanced Pancreatic Cancer With Opioid Growth Factor (OGF): Phase 2
Rationale: Opioid growth factor may stop the growth of pancreatic cancer by blocking blood flow to the tumor
Primary Outcome Measures: This phase 2 trial is studying how well opioid growth factor works in treating patients with advanced pancreatic cancer that cannot be removed by surgery

CROHN'S DISEASE

Study Type: Human, open-labeled, pilot prospective trial, twelve weeks
Subjects: Seventeen, mean CDAI score of 356 +/- 27

Medication: 4.5 mg naltrexone/day, evening administration
Where: Pennsylvania State University College of Medicine
Researchers: Jill P. Smith and colleagues
Status: Completed
Website: http://www.hmc.psu.edu/colorectal/research/naltrexone.htm
Contact: Sandra Bingaman, RN; Phone: 717-531-8108
Presented: "Low-Dose Naltrexone as a Treatment For Active Crohn's Disease,"
                May 23, 2006, Digestive Diseases Week Conference
Published: Jill Smith, Heather Stock, Sandra Bingaman, David Mauger, Moshe
                Rogsnizky, and Ian Zagon, "Low-Dose Naltrexone Therapy Improves
                Active Crohn's Disease," *American Journal of Gastroenterology* 102, no.4
                (2007), 1–9
Results: CDAI scores decreased significantly (P= 0.01) with LDN, and remained
                lower than baseline four weeks after completing therapy; 89% of pa-
                tients exhibited a response to therapy and 67% achieved a remission (P
                < 0.001); improvement was recorded in both quality of life surveys with
                LDN compared with baseline
Side Effects: No laboratory abnormalities were noted; most common side effect was
                sleep disturbances, occurring in seven patients
Conclusion: LDN therapy appears effective and safe in subjects with active Crohn's
                disease; further studies are needed to explore the use of this compound

CROHN'S DISEASE

Study Type: Human, phase 2, placebo-controlled clinical trial
Subjects: Forty
Medication: 4.5 mg naltrexone/day
Where: Pennsylvania State University College of Medicine
Researchers: Jill P. Smith and colleagues
Status: In progress; recruiting began mid 2007, twenty subjects enrolled as of
                March 2008
Website: http://www.hmc.psu.edu/colorectal/research/naltrexone.htm
Contact: Sandra Bingaman, RN, Phone: 717-531-8108

HIV, AIDS

Study Type: Human; nine-month study involves three study groups: LDN treatment
                only, LDN plus antiretroviral drugs, and only antiretroviral drugs.
Status: Ongoing; start date September 2007; still enrolling as of February 2008
Subjects: 250 adults
Medication: LDN
Where: University Hospital in Bamako, Mali
Researchers: Jaquelyn McCandless, Abdel Kader Traore
Website: http://www.ldnafricaaids.org
Contact: Jaquelyn McCandless, E-mail: jmccandless@prodigy.net
Presented: Ongoing status available at http://www.ldnafricaaids.org
Official Title: The Mali HIV+ AIDS LDN Initiative
Rationale: The safety as well as potential efficacy of LDN in preventing AIDS was
                discovered by Bernard Bihari, M.D., in 1985; since that time Dr. Bihari
                has treated more than 350 patients, 94% of whom have remained HIV

positive without progression into AIDS for up to eighteen or more years so far

Outcome Measures:
: Subjects must have reduced CD4 counts in the 275 to 475 cells range at the outset; laboratory studies will be rechecked at twelve-week intervals.

Purpose:
: Prevent progression of HIV into AIDS

AIDS

Study Type:
: Human; three-month, double-blind, placebo-controlled, crossover

Status:
: Completed 1986

Subjects:
: Thirty-eight patients with AIDS; twenty-two randomly assigned to naltrexone and sixteen to placebo; after three months the placebo patients were switched to naltrexone

Medication:
: 1.75 mg of oral LDN administered at bedtime

Where:
: Downstate Medical Center, Brooklyn, NY

Researchers:
: Bernard Bihari, Finvola M. Drury, Vincent P. Ragone, et al.

Website:
: http://www.lowdosenaltrexone.org/ldn_aids_1988.htm

Contact:
: email@lowdosenaltrexone.org

Presented:
: International AIDS Conference in Stockholm, Sweden, 1988

Official Title:
: Low Dose Naltrexone in the Treatment of Acquired Immune Deficiency Syndrome

Rationale:
: In the light of recent evidence suggesting that the endorphinergic system plays an important role in the homeostatic regulation of immune function, we have developed and tested an immunoenhancing treatment approach using low dose naltrexone; the information already available about the physiology and pathophysiology of alpha IFN suggests that its reduction is the crucial factor in patient stabilization in our study

Outcome Measures:
: Decrease in serum alpha IFN levels, number of opportunistic infections, patient stabilization.

Purpose:
: Assess the effect of low dose naltrexone as an immunoenhancing agent in the treatment of AIDS

Side Effects:
: No significant adverse reactions.

Results:
: The results indicate that 61% (23/38) of AIDS patients in this study treated with low dose naltrexone respond with a marked decrease in serum alpha IFN levels. This decrease appears to result in changes in immune function that provide relative protection to this group from progression and death (in the time frame of this study) and/or is a marker for such protective changes; the clinical protection has not yet been accompanied by a rise in T4 absolute numbers or in the T4/T8

Conclusion:
: Our results reinforce the possibility that the serum alpha IFN level is a good marker for following the course of AIDS and for assessing patient response to new treatment approaches; our results also suggest that further studies of endorphinergic function and dysfunction in AIDS may yield an expanded understanding of the pathophysiology of the disease and development of effective immune modulating treatment approaches

IRRITABLE BOWEL SYNDROME (IBS)

Study Type:	Human, open-labeled, four weeks
# Subjects:	Forty-two, with IBS diagnosis based on Rome II criteria
Medication:	PTI-901 low-dose naltrexone (0.5 mg daily)
Where:	Department of Gastrointestinal and Liver Diseases, Tel-Aviv University, Tel-Aviv, Israel
Researchers:	R. Kariv and colleagues
Status:	Completed
Website:	http://www.ncbi.nlm.nih.gov/pubmed/17080248
Published:	Kariv R, Tiomny E, Grenshpon R, et al, Low-dose naltrexone for the treatment of irritable bowel syndrome: a pilot study, Digestive Diseases Sciences, 2006 Dec;51(12):2128–33.
Official Title:	Low-dose Naltrexone for the Treatment of Irritable Bowel Syndrome: A Pilot Study
Rationale:	Preclinical studies have shown that a very low dose of naltrexone hydrochloride (NTX), an opiate antagonist, can block excitatory opioid receptors without affecting inhibitory opioid receptors, resulting in analgesic potency without side effects
Primary Outcome Measures:	Number of pain-free days and overall symptom relief, evaluated by a global assessment score
Purpose:	Assessed the efficacy and safety of PTI-901 (low dose naltrexone hydrochloride) treatment in irritable bowel syndrome (IBS) patients
Side Effects:	No significant adverse reactions
Results:	During treatment, the mean weekly number of pain-free days increased from 0.5+/-1 to 1.25+/-2.14 (P=0.011)
Conclusion:	PTI-901(low dose naltrexone) improves pain and overall feeling, and is well tolerated by IBS patients; a large, randomized, double-blind, placebo-controlled study is justified

IRRITABLE BOWEL SYNDROME

Study Type:	Human, phase 3, randomized, double-blind, multi-center
# Subjects:	600 women with documented IBS
Medication:	PTI-901 lowdose naltrexone (0.5 mg daily)
Where:	Pain Therapeutics, Inc., San Francisco
Researchers:	Pain Therapeutics, Inc.
Status:	Completed
Website:	http://www.paintrials.com
Contact:	Pain Therapeutics, Inc., 650-624-8200
Published:	Kariv, R., E. Tiomny, R. Grenshpon, et al., "Low-dose Naltrexone for the Treatment of Irritable Bowel Syndrome: A Pilot Study," Digestive Diseases Sciences 51, no. 12 (December 2006), 2128-33.
Official Title:	Low-dose Naltrexone for the Treatment of Irritable Bowel Syndrome: A Pilot Study
Rationale:	Successful pilot study at Tel-Aviv University with low dose naltrexone for IBS
Primary Outcome Measures:	Meaningful relief of IBS symptoms in the third month of treatment
Purpose:	Assess the efficacy PTI-901 (low dose naltrexone) treatment in irritable bowel syndrome (IBS) patients

Side Effects: Favorable safety profile
Results: Patients reported statistically meaningful relief of IBS symptoms in the second month of treatment ($p < 0.02$), but the drug did not demonstrate a meaningful benefit in the third month of treatment, which was defined as the primary endpoint
Conclusion: As of December 2005, the company discontinued all further clinical development activities with PTI-901 (low dose naltrexone)
Note: *The LDN patient and practitioner community has observed that naltrexone at the low levels used in the pain therapeutics trials (0.5 mg daily) is not high enough to be an effective dose level for treating IBS, or any other condition for which LDN has been found to be beneficial. Many patients have reported success for gastrointestinal conditions such as IBS, ulcerative colitis and Crohn's at the therapeutic level of 4.5 mg daily. See the Crohn's disease trials section.*

FIBROMYALGIA

Condition: Primary fibromyalgia
Study Type: Human; twelve-week placebo-controlled, single-blind, pre-post drug trial
Status: Ongoing; start date June 2007; still enrolling as of February 2008
Subjects: Ten
Medication: Low dose naltrexone
Where: Stanford University, Stanford Systems Neuroscience and Pain Lab
Researchers: Jarred Younger and Sean Mackey
Website: http://snapl.stanford.edu/ldn
Contact: Jarred Younger, E-mail: LDN_Younger@stanford.edu
Official Title: Effects of Low Dose Naltrexone in Fibromyalgia
Rationale: Low dose naltrexone (LDN) has been reported anecdotally to reduce the symptoms of fibromyalgia; the drug may work by regulating natural pain-reducing systems
Outcome   pain, fatigue, sleep quality
Measures:
Primary:
Secondary: mechanical pain sensitivity, thermal pain sensitivity
Purpose: In this study, both LDN and placebo will be administered to a small group of individuals with fibromyalgia to assess the drug's efficacy in treating the condition

MULTIPLE SCLEROSIS

Condition: Multiple sclerosis (not distinguished by type)
Study Type: Human; randomized, placebo-controlled, crossover, parallel-group study
Status: Completed November 2007, awaiting presentation and publication
Subjects: Eighty
Medication: 4.5 mg naltrexone/day, evening administration
Where: MS Center, University of California, San Francisco
Researchers: Bruce Cree and colleagues
Website: http://ucsf.edu/msc/research.htm
Contact: Phone: 415-353-2069
Presentation: Pending positive results, American Academy of Neurology, April 2008

Primary Outcome Measures:	Effects of low dose naltrexone on the multiple sclerosis quality of life inventory (MSQLI54)

Condition:	Multiple sclerosis (PPMS, according to McDonald criteria)
Study Type:	Human; six-month, pilot, multicentric, open-label, therapeutic study
Status:	Completed December 2007, awaiting presentation and publication
# Subjects:	Forty
Medication:	Low dose naltrexone
Where:	Neurology Department, University of Milan, Italy
Researchers:	Maira Gironi, and colleagues
Presentation	Four-month data presented at the European Congress of MS in Prague, October 2007; final six-months results at American Academy of Neurology, April 2008
Official Title:	Pilot Multicentric Study of Low Dose Naltrexone in Primary Progressive Multiple Sclerosis
Primary Outcome Measures:	Clinical and biochemical evaluations; beta endorphin blood levels, evaluation of LDN safety and efficacy on spasticity, pain and fatigue
Side Effects at Four Months:	Transitory hematological abnormalities (increase of liver enzymes, hypercholesterolemia), mild agitation and sleep disturbance were the commonest adverse events; only two dropouts occurred (one for protocol violation and one for severe increase of hypertonia)
Conclusion at	LDN has shown, so far, to be a safe treatment in PPMS; improves quality of life; Phase 2 trials recommended

EXPERIMENTAL AUTOIMMUNE ENCEPHALOMYELITIS (EAE; A MODEL THAT MIMICS MS)

Study Type:	Animal
Status:	Completed August 2007; awaiting publication
Medication:	This study will be treating an animal model of MS daily with either a high dose of naltrexone or a low dose of naltrexone to determine whether naltrexone influences disease course
Where:	Pennsylvania State University College of Medicine
Researchers:	Ian S. Zagon and colleagues
Website:	http://fred.psu.edu/ds/retrieve/fred/investigator/isz1
Contact:	Phone: 717-531-8650
Official Title:	Role of Opioid Peptides and Receptors in MS
Rationale:	Evidence for the involvement of endogenous opioids and opioid receptors in MS will open a new field of research related to the pathogenesis of this disease, and contribute to the development of strategies for treatment
Primary Outcome Measures:	Expectations are that continuous opioid receptor blockade will exacerbate the progression of MS, whereas a low dose of naltrexone will retard the course of this disease

MULTIPLE SCLEROSIS (NOT DISTINGUISHED BY TYPE)

Study Type:	Human; sixteen-week, double-blind, randomized, placebo-controlled, crossover-design

Status: Pending funding as of March 2008
Subjects: Thirty-six
Medication: Low dose naltrexone
Where: Summa Health System, Akron, Ohio
Researchers: David Pincus
Website: http://www.mindbrainconsortium.com
Contact: Phone: 330-379-TMBC
Primary   Study will examine symptom severity as well as any changes in quality
Outcome   of life, sleep patterns, and affective states
Measures

AUTOIMMUNE DISEASES (UVEITIS, BEHÇET'S DISEASE, ALLERGIC CONJUNCTIVITIS/RHINITIS, OPTIC NEURITIS, MULTIPLE SCLEROSIS, SLE/OVERLAP SYNDROME, RHEUMATOID ARTHRITIS, EVANS SYNDROME/HEMOLYTIC ANEMIA)

Study Type: Human; observational
Status: Completed 1999
Subjects: Thirty-eight patients
Medication: Met-enkephalin (peptid-M, LUPEX) was administered s.c. dissolved in
                 2 ml of physiological saline, 0.2–0.3 mg/kg five times weekly for four
                 weeks. Then weekly dose gradually reduced every four weeks till 5 mg
                 weekly
Where: Croatian Institute of Brain Research, Zagreb University School of Med-
                 icine, Department of Neurology, Zagreb, Croatia
Researchers: N. Stambuk, V. Brinar, Z. Brzovic, et al.
Contact: Nikola Stambuk, Rugjer Boskovic Institute, Bijenicka 54, HR-10001 Za-
                 greb, Croatia, E-mail: stambuk@rudjer.irb.hr
Presented: 2nd International Congress of Autoimmunity, Tel Aviv, March 7, 1999
Published: Stambuk, N., V. Brinar, Z. Brzovic, et al., "Met-enkephalin Therapy for
                 Autoimmune Diseases: Selective Immunomodulation and Extention of
                 Steroid Therapy," *Journal of Autoimmunity*, Sppl. *Abstracts of 2nd In-
                 ternational Congress of Autoimmunity*, edited by J. F. Bach. London:
                 Academic Press, 1999, 86.
Official Title: Met-enkephalin Therapy for Autoimmune Diseases: Selective Im-
                 munomodulation and Extension of Steroid Therapy
Rationale: Met-enkephalin is a pentapeptide present in different tissues; this neu-
                 ropeptide exerts various modulatory signals on different cell types,
                 which led to its application in clinical medicine
Outcome   Clinical parameters including EDSS, evoked potentials, visual field, MRI
Measures: and ultrasonographic findings were observed and evaluated
Purpose: Define treatment protocols for combined immunotherapy of Met-
                 enkephalin and corticosteroids or i.v. IgG (7S, 5S+7S)
Side Effects: No toxicity or side effects were observed during the Met-enkephalin
                 therapy ranging from six months to over three years
Results: During Met-enkephalin therapy, relapses were rare, mild, and in most
                 cases did not require additional corticosteroid therapy, that is, only the
                 rise of Met-enkephalin dose was sufficient; however, the synergistic ac-
                 tion of Met-enkephalin and corticosteroids is discussed considering the
                 fact that it is a part of ACTH precursor molecule; in addition to the
                 beneficial clinical effects and significant reduction in the number and

severity of relapses, the peptide modulated serum IFN, IFN-gamma, phagocyte, NK and lymphocyte functions

Conclusion: This study emphasizes that in severe autoimmune diseases parallel administration of Met-enkephalin and corticosteroids enables significant reduction of the steroid dose (to 2–4 mg dexamethasone daily, for example, during relapse); finally, we reevaluate classic opioid hypothesis of Met-enkephalin action considering Met-enkephalin interactions with its probable calpastatin receptor site, that was recently defined by means of the Molecular Recognition Theory

OPIOID INDUCED CONSTIPATION (OIC)

Study Type: Human; double-blind, randomized, placebo-controlled trial

Subjects: Twenty-two (nine men and thirteen women) enrolled in a methadone maintenance program and having methadone-induced constipation

Medication: Low-dose methylnaltrexone up to 0.365 mg/kg

Where: Pritzker School of Medicine, University of Chicago

Researchers: Chun-Su Yuan, Joseph F. Foss, and colleagues

Status: Completed

Contact: Chun-Su Yuan, E-mail: cyuan@uchicago.edu

Published: Yuan, Chun-Su, Joseph F. Foss, et al., "Methylnaltrexone for Reversal of Constipation Due to Chronic Methadone Use," Journal of the American Medical Association 283 (2000), 367–372, http://jama.ama-assn.org/cgi/content/full/283/3/367

Official Title: Methylnaltrexone for Reversal of Constipation Due to Chronic Methadone Use

Rationale: Constipation is the most common chronic adverse effect of opioid pain medications in patients who require long-term opioid administration, such as patients with advanced cancer, but conventional measures for ameliorating constipation often are insufficient

Primary   Laxation response, oral-cecal transit time, and central opioid with-
Outcome   drawal symptoms were compared between the placebo and drug groups
Measures:

Purpose: To evaluate the efficacy of methylnaltrexone, the first peripheral opioid receptor antagonist, in treating chronic methadone-induced constipation

Side Effects: No opioid withdrawal was observed in any subject, and no significant adverse effects were reported by the subjects during the study

Results: The eleven subjects in the placebo group showed no laxation response, and all eleven subjects in the intervention group had laxation response after intravenous methylnaltrexone administration (P<.001). The oral-cecal transit times at baseline for subjects in the methylnaltrexone and placebo groups averaged 132.3 and 126.8 minutes, respectively. The average (SD) change in the methylnaltrexone-treated group was -77.7 (37.2) minutes, significantly greater than the average change in the placebo group (-1.4 [12.0] minutes; P<.001)

Conclusion: Low-dosage methylnaltrexone may have clinical utility in managing opioid-induced constipation

Resources

Books

Aisen, Paul, Deborah Marin, and Kenneth Davis. *Inflammatory Processes—Anti-Inflammatory Therapy, in Alzheimer Disease: From Molecular Biology to Therapy.* Boston: Birkhauser Publishing, 1996.

Antel, Jack, Gary Birnbaum, Hans-Peter Hartung, and Angela Vincent, *Clinical Neuroimmunology*, New York: Oxford University Press, 2005.

Russell Blaylock. *Excitotoxins: The Taste that Kills.* Santa Fe, NM: Health Press, 1994.

Bradley, Mary. *Up the Creek with a Paddle: Beat MS and Many Autoimmune Disorders with Low Dose Naltrexone*, Frederick, MD: Publish America, 2005.

Daruna, Jorge H. *Introduction to Psychoneuroimmunology.* Boston: Elsevier Academic Press, 2004.

Demetrios, Julius, and Pierre Renault, ed. *Narcotic Antagonists, Naltrexone: Progress Report*, NIDA Research Monograph 9. Rockville, MD: U.S. Department of Health, Education and Welfare, 1976.

Galoyan, Armen, Hugo Besedovsky, and Abel Lajtha. *Handbook of Neurochemistry and Molecular Neurobiology: Neuroimmunology*, 3rd ed. New York: Springer, 2008.

Goodman, Louis Sanford, Alfred Gilman, Laurence L. Brunton, John S Lazo, and Keith L. Parker, eds. *Goodman & Gilman's The Pharmacological Basis of Therapeutics*, 11th edition. New York: McGraw-Hill, 2006.

Guilmette, Bruce. *There's More to Life Than Just Living: A Personal Story about Cancer Survival.* Mustang, OK: Tate Publishing, 2007.

Kalman, Bernadette, and Thomas H. Brannagan III. *Neuroimmunology in Clinical Practice.* Malden, MA: Blackwell, 2007.

McCandless, Jaquelyn. *Children with Starving Brains: A Medical Treatment Guide for Autism Spectrum Disorder*, 3rd ed. Putney, VT: Bramble Books, 2007.

Moore, Elaine A. *Autoimmune Diseases and Their Environmental Triggers.* Jefferson, NC: McFarland, 2002.

Pert, Candace B. *Molecules of Emotion: The Science Behind Mind-Body Medicine.* New York: Simon & Schuster, 1999.

Sacerdote, Paola, Elena Limironi, and Leda Gaspani. *Experimental Evidence for Immunomodulatory Effects of Opioids*, Landes Bioscience, 2000–2005, http://www.ncbi.nlm.nih.gov/books/bv.fcgi?rid=eurekah.section.10979 (accessed May 21, 2008).

Singleton, Ken. *The Lyme Disease Solution.* South Lake Tahoe, CA: BioMed Publishing Group, 2008.

Naltrexone, LDN, and OGF Reports and Research

"Best-Case Series for the Use of Immuno-Augmentation Therapy and Naltrexone for the

Treatment of Cancer." Agency for Healthcare Research and Quality, Evidence Report/ Technology Assessment Number 78, April 2003, http://www.ahrq.gov/clinic/epcsums /immaugsum.htm (accessed May 14, 2008).

Braude, Monique, and J. Michael Morrison. "Preclinical Toxicity Studies of Naltrexone." In *Narcotic Antagonists, Naltrexone: Progress Report*, edited by Julius Demetrios and Pierre Renault, NIDA Research Monograph 9. Rockville, MD: U.S. Department of Health, Education and Welfare, 1976.

Gluck, David. Introductory Overview, NCI Conference: "Low Dose Opioid Blockers, Endorphins and Metenkephalins Promising Compounds for Unmet Medical Needs." Bethesda, Maryland, April 20, 2007.

"Opioid Growth Factor." NCI Drug Dictionary, National Cancer Institute. http:// www.cancer.gov/Templates/drugdictionary.aspx/?CdrID=428488 (accessed May 14, 2008).

Plotnikoff, Nicholas. "Overview of Metenkephalin Actions." NCI Conference: "Low Dose Opioid Blockers, Endorphins and Metenkephalins: Promising Compounds for Unmet Medical Needs," Bethesda, Maryland, April 20, 2007.

Smith, Jill. "Animal Studies with Naltrexone." NCI Conference: "Low Dose Opioid Blockers, Endorphins and Metenkephalins: Promising Compounds for Unmet Medical Needs," Bethesda, Maryland, April 20, 2007.

Autism Spectrum Disorder Reports and Research

"Autism Fact Sheet," National Institute of Neurological Disorders and Stroke, http:// www.ninds.nih.gov/disorders/autism/detail_autism.htm (accessed May 18, 2008).

Bouvard, M. P., M. Leboyer, J. Launay, C. Rescasens, M. Plumet, D. Waller-Perotte, F. Tabuteau, et al. "Low-dose Naltrexone Effects on Plasma Chemistries and Clinical Symptoms in Autism: A Double-blind, Placebo-controlled Study." *Psychiatry Research* 58 no. 3 (October 1995), 191–201.

Campbell, M. "Naltrexone in Autistic Children: Behavioral Symptoms and Attentional Learning." *Journal of the American Academy of Child and Adolescent Psychiatry* 32, no. 6 (1993), 1283–91.

Edelson, Stephen M. "Naltrexone and Autism." Autism Research Institute. www. autism.org/naltrex.html (accessed May 21, 2008).

"Interview with Professor Jaak Panksepp." Autism Research Institute, March 11, 1997. http://www.autism.org/interview/panksepp.html (accessed May 18, 2008).

Leboyer, M., P. Bouvard, J. Launay, C. Rescasens, M. Plumet, D. Waller-Perotte, "Opiate Hypothesis in Infantile Autism? Therapeutic Trials with Naltrexone." *Encephale* 19 no. 2 (March–April 1993), 95–102.

McCandless, Jaquelyn. "Autism, Autoimmunity, and Low-dose Naltrexone." Presented at the 2nd Annual LDN Conference, April 7, 2006, National Library of Medicine, NIH Campus, Bethesda, Maryland. http://www.lowdosenaltrexone.org/_conf2006/J_ McCandless.pdf (accessed May 21, 2008).

_____. "Low-Dose Naltrexone (LDN) for Mood Regulation and Immunomodulation in ASD." lowdosenaltrexone.org, http://www.lowdosenaltrexone.org/_conf2006/J _McCandless2.pdf (accessed May 21, 2008).

Sahley, T. L. and J. Panksepp. "Brain Opioids and Autism: An Updated Analysis of Possible Linkages." *Journal of Autism and Developmental Disorders* 17, no. 2 (June, 1987), 201–16.

Cancer Reports and Research

Berkson, Burton. "Clinical Experience with Low Dose Naltrexone Protocols for Various Cancers." NCI Conference: "Low Dose Opioid Blockers, Endorphins and Metenkephalins: Promising Compounds for Unmet Medical Needs," Bethesda, Maryland, April 20, 2007. http://www.lowdosenaltrexone.org/_mm/Berkson_Presentation_Apr_2007.pdf (accessed May 21, 2008).

Berkson, B., D. Rubin, and A. Berkson. "The Long-Term Survival of a Patient with Pancreatic Cancer with Metastases to the Liver after Treatment with the (-Lipoic Acid/ Low Dose Naltrexone Treatment Protocol." *Integrative Cancer Therapies* 5 no. 1 (2006), 83–89.

_____. "Reversal of Signs and Symptoms of a B-cell Lymphoma in a Patient Using Only Low-dose Naltrexone." *Integrative Cancer Therapies* 6 no. 3 (September 2007), 293–296. http://www.ldn4cancer.com/files/berkson-b-cell-lymphoma-paper.pdf (accessed May 14, 2008).

Bisignani, G. J., P. J. McLaughlin, S. D. Ordille, M. S. Beltz, M. V. Jarowenko, and I. S. Zagon, "Human Renal Cell Cancer Proliferation in Tissue Culture is Tonically Inhibited by Opioid Growth Factor." *Journal of Urology* 162, no. 6 (December 1999), 2186–91.

Blebea, J., J. Mazo, T. Kihari, J. H. Vu, P. J. McLaughlin, R. G. Atnip, and I. S. Zagon. "Opioid Growth Factor Modulates Angiogenesis." *Journal of Vascular Surgery* 32 no. 2 (August 2000), 364–373.

_____, J.-H. Vu, S. Assadnia, P. J. McLaughlin, R. G. Atnip, and I. S. Zagon. "Differential Effects of Vascular Growth Factors on Arterial and Venous Angiogenesis." *Journal of Vascular Surgery* 35 (2002), 532–538.

Cheng, Fan, Patricia McLaughlin, Michael Verderame, and Ian S. Zagon. "The OGF-OGFr Axis Utilizes the p21 Pathway to Restrict Progression of Human Pancreatic Cancer." *Molecular Cancer*, January 11, 2008. Provisional abstract available online at http://www.molecular-cancer.com/content/7/1/5 (accessed May 14, 2008).

_____, Ian S. Zagon, Michael F. Verderame, and Patricia McLaughlin. "The Opioid Growth Factor (OGF)-OGF Receptor Axis Uses the p16 Pathway to Inhibit Head and Neck Cancer." *Cancer Research* 67 (November 1, 2007), 10511–10518.

Jaglowski, J.R., I. S. Zagon, B. C. Stack Jr., M. F. Verderame, A. E. Leure-duPree, J. D. Manning, and P. J. McLaughlin. "Opioid Growth Factor Enhances Tumor Growth Inhibition and Increases the survival of Paclitaxel-treated Mice with Squamous Cell Carcinoma of the Head and Neck." *Cancer Chemotherapy and Pharmacology* 56 no. 1 (July 2005), 97–104.

Kampa M., E. Bakogeorgou, A. Hatzoglou, et al. "Opioid Alkaloids and Casomorphin Peptides Decrease the Proliferation of Prostatic Cancer Cell Lines (LNCaP, PC3 and DU145) through a Partial Interaction with Opioid Receptors." *European Journal of Pharmacology* 335, no. 2–3 (September 24, 1997), 255–65.

Lissoni, Paolo, Fabio Malugani, Ola Malysheva, Vladimir Kozlov, Moshe Laudon, Ario Conti, and Georges Maestroni., "Neuroimmunotherapy of Untreatable Metastatic Solid Tumors with Subcutaneous Low-dose Interleukin-2, Melatonin and Naltrexone. *Neuroendocrinology Letters* 23, no. 4 (August 200), 341–344, http://www.nel.edu/23_4/NEL230402A09_Lissoni.htm (accessed May 14, 2008).

McLaughlin, P. J., B. C. Stack Jr., R. J. Levin, F. Fedok, I. S. Zagon, "Defects in the Opioid Growth Factor Receptor in Human Squamous Cell Carcinoma of the Head and Neck," *Cancer* 197 no. 7 (April 2003),1701–1710.

_____, and I. S. Zagon. "Modulation of Human Neuroblastoma Transplanted into Nude Mice by Endogenous Opioid Systems." *Life Sciences* 41, no. 12 (September 1987), 1465–72.

"Opioid Growth Factor (OGF) in Cancer Therapy, A Unique Biotherapeutic Agent,"
    MedInsight Research Institute, October 2006, (http://www.medvision.com/mihpf/
    medinsight%20-%20ogf%20review.pdf, accessed May 2008).
Plotnikoff N. P. "Opioids: Immunomodulators: A Proposed Role in Cancer and Aging."
    Annals of New York Academy of Sciences 521 (1988), 312–22.
Plotnikoff, N.P., G. C. Miller, N. Nimeh, et al. "Enkephalins and T-cell Enhancement
    in Normal Volunteers and Cancer Patients." Annals of New York Academy of Sciences
    496 (1987), 608–19.
Smith, J. P., R. I. Conter, S. I. Bingaman, H. A. Harvey, D. T. Mauger, M. Ahmad, L.
    M. Demers, W. B. Stanley, P. J. McLaughlin, and I. S. Zagon. "Treatment of Advanced
    Pancreatic Cancer with Opioid Growth Factor: Phase I." Anticancer Drugs 15 no. 3
    (March 2004), 203–209.
Zagon I. S., S. D. Hytrek, C. M. Lang, J. P. Smith, et al. "Opioid Growth Factor
    ([Met5]enkephalin) Prevents the Incidence and Retards the Growth of Human Colon
    Cancer." American Journal of Physiology 271, no.3 pt. 2 (September 1996), R780–6.
_____, _____, and P. J. McLaughlin. "Opioid Growth Factor Tonically Inhibits Human
    Colon Cancer Cell Proliferation in Tissue Culture. American Journal of Regulatory,
    Integrative and Comparative Physiology 271, no. 30 (1996), 411–518.
_____, Jeffrey R. Jaglowski, Michael F. Verderame, Jill P. Smith, Alphonse E. Leure-
    duPree, and Patricia J. McLaughlin. "Combination Chemotherapy with Gemcitabine
    and Biotherapy with Opioid Growth Factor (OGF) Enhances the Growth Inhibition
    of Pancreatic Adenocarcinoma." Cancer Chemotherapy and Pharmacology 5 no. 5
    (November 2005), 510–520.
_____, and P. McLaughlin. "Endogenous Opioids and the Growth Regulation of a
    Neural Tumor." Life Sciences 43, no. 16 (1988),1313–8.
_____. "Naltrexone Modulates Tumor Response in Mice with Neuroblastoma." Science
    221, no. 4611 (August 12, 1983), 671–673.
_____. "Opioid Growth Factor Receptor is Unaltered with the Progression of Human
    Pancreatic and Colon Cancers." International Journal of Oncology 29, no. 2 (August
    2006), 489–494.
_____, "Opioids and Differentiation in Human Cancer Cells." Neuropeptides 39, no. 5
    (October, 2005), 495–505.
_____, P. J .McLaughlin, S. R. Goodman, and R. E. Rhodes. "Opioid Receptors and En-
    dogenous Opioids in Diverse Human and Animal Cancers." Journal National Can-
    cer Institute 79, no. 5 (1987),1059–65.
_____, C. D. Roesner, M. F. Verderame, B. M. Olsson-Wilhelm, R. J. Levin, and P. J.
    McLaughlin. "Opioid Growth Factor Regulates the Cell Cycle of Human Neoplasias."
    International Journal of Oncology 17, no. 5 (November 2000), 1053–1061.
_____ [Ian], M. F. Verderame, J. Hankins, and P. J. McLaughlin. "Overexpression of
    the Opioid Growth Factor Receptor Potentiates Growth Inhibition in Human Pan-
    creatic Cancer." International Journal of Oncology 30 no. 4 (April 2007), 775–783.

Crohn's Disease Reports and Research

Konjevoda, P. et al. "Cytoprotective Effects of Met-enkephalin and Alpha-MSH on
    Ethanol Induced Gastric Lesions in Rats." Journal of Physiology—Paris 95 (2001),
    277–281.
"MedInsight Announces Clinical Trial Results of Low-dose Naltrexone: Potential Break-
    through for Crohn's Disease Patients." Medical News Today, January 31, 2007.
    http://wwww.medicalnewstoday.com/articles/61960.php (accessed May 13, 2008).
Smith, Jill, Heather Stock, Sandra Bingaman, David Mauger, Moshe Rogonitzky, and

Ian Zagon. "Low-Dose Naltrexone Therapy Improves Active Crohn's Disease." *American Journal of Gastroenterology* 102, no. 4 (2007), 1–9.

Immune System and HIV/AIDS Reports and Research

Bihari, Bernard, Finvola Drury, Vincent Ragone, Gennaro Ottomanelli, Elena Buimovici-Leine, Milagros Orbe, William Foeste, Johnny Thomas, and Robert Kirt. "Low Dose Naltrexone in the Treatment of Acquired Immune Deficiency Syndrome." Oral Presentation at the IV International AIDS Conference in Stockholm, June 1988, lowdosenaltexone.org, http://www.lowdosenaltrexone.org/ldn_aids_1988.htm (accessed May 20, 2008).

Bihari, B., and N. P. Plotnikoff. "Methionine Enkephalin in the Treatment of AIDS-Related Complex." In *Cytokines: Stress & Immunity*, edited by N. P. Plotnikoff, A. Murgo, R. E. Faith, and R. A. Good. Boca Raton, FL: CRC Press, 1999.

"Dr. Kamau B. Kokayi Interviews Dr. Bihari, September 23, 2003, WBAI in New York City, *Global Medicine Review*," http://www.gazorpa.com/interview.html (accessed July 10, 2007).

Gekker, G., J. Lokensgard, P. Peterson, "Naltrexone Potentiates Anti-HIV-2 Activity of Antiretroviral Drugs in CD4+ Lymphocyte Cultures, *Drug and Alcohol Dependence* 64, no. 3 (November 2001), 257–263.

"Low-dose Naltrexone and Lipodystrophy," The Body, Seattle Treatment Education Project, May 19, 2000, http://www.thebody.com/content/art2019.html (accessed October 15, 2007).

"Naltrexone—A Unique Immune Regulator Showing Promise in All Manner of Autoimmune Diseases Including HIV," International Aging Systems, http://www.antiagingsystems.com/a2z//naltrexone.htm (accessed July 29, 2007).

Plotnikoff, Nicholas. "Methionine Enkephalin, A New Cytokine with Antiviral and Anti-tumor Properties," in *Potentiating Health and the Crisis of the Immune System*, edited by A. Mizrahi, Stephen Fulder, and Nimrod Sheinman, chapter 19. New York: Springer, 1997.

_____, and J. Wybran. "Methionine-enkephalin Shows Promise in Reducing HIV in Blood." *American Family Physician* 40, no. 3 (September 1989), 234.

Wybran, J., and N. P. Plotnikoff. "Methionine-Enkephalin: A New Lymphokine for the Treatment of ARC Patients." In *Stress and Immunity*, edited by N. J. Plotnikoff, A. J. Murgo, R. Faith, and J. Wybran. Boca Raton, FL:: CRC Press, 1991.

_____, L. Schandene, J. P. Van Vooren, et al. "Immunologic Properties of Methionine-enkephalin, and I in AIDS, ARC, and Cancer." *Annals of New York Academy of Sciences* 496 (1987), 108–14.

Irritable Bowel Syndrome Reports and Research

Generalia, Joyce, and Dennis Cada. "Naltrexone: Irritable Bowel Syndrome." *Hospital Pharmacy* 42, no. 8 (August 2007), 712–718.

Kariv, R., E. Tiomny, R. Grenshpon, R. Dekel, G. Wasiman, Y. Ringel, and Z. Halpern. "Low-dose Naltrexone for the Treatment of Irritable Bowel Syndrome: A Pilot Study." *Digestive Diseases Sciences* 51, no. 12 (December 2006), 2128–2133.

Keren, David, and James Goeken. "Autoimmune Reactivity in Inflammatory Bowel Diseases." In *Progress and Controversies in Autoimmune Disease Testing*, edited by David F. Keren and Robert M. Nakamura. Philadelphia: W. B. Saunders, 1997, 465–481.

Multiple Sclerosis Reports and Research

Agrawal, Yash. "Low Dose Naltrexone Therapy in Multiple Sclerosis. *Medical Hypotheses* 64, no. 4 (2005), 721–724.

_____. "The Need for Trials of Low Dose Naltrexone as a Possible Therapy for Multiple Sclerosis." Interviewed with Rob Lester of the Boston Cure Project, January 2005 (http://www.acceleratedcure.org/downloads/interview-agrawal.pdf, accessed May 13, 2008).

_____. "Possible Importance of Antibiotics and Naltrexone in Neurodegenerative Disease." Letter to the Editor, *European Journal of Neurology* 13, no. 9 (2006), e7.

"Can Naltrexone Relieve MS Symptoms." *New Horizons* (Brewer Science Library, Richland Center, Wisconsin), Winter 1999. http://www.mwt.net/~drbrewer/naltrexms. htm (accessed May 13, 2008).

Chaudhuri, A., and P. Behan. "Multiple Sclerosis Is Not an Autoimmune Disease." *Archives of Neurology* 61, no. 10 (October 2004), 1610–1612.

Gironi, M., R. Furlan, M. Rovaris, G. Comi, M. Filippi, A. Panerai, and P. Sacerdote. "Beta Endorphin Concentrations in PBMC of Patients with Different Clinical Phenotypes of Multiple Sclerosis." *Journal of Neurology, Neurosurgery & Psychiatry* 74, no. 4 (April 2003), 495–97.

_____, V. Martinelli, E. Brambilla, R. Furlan, A. E. Panerai, G. Comi, P. Sacerdote. "Beta-endorphin Concentrations in Peripheral Blood Mononuclear Cells of Patients with Multiple Sclerosis: Effects of Treatment with Interferon Beta." *Archives of Neurology* 57, no. 8, (August 2000),1178–81.

Grigoriadis, Nikolaos, and Georgios Hadjigeorgious. "Virus-mediated Autoimmunity in Multiple Sclerosis." *Journal of Autoimmune Diseases* 3, no. 1 (February 2006), http://www.jautoimdis.com/content/3/1/1 (accessed May 13, 2008).

Martino, G., R. Furlan, E. Brambilla, A. Bergami, F. Ruffini, M. Gironi, P. L. Poliani, L. M. Grimaldi, and G. Comi. "Cytokines and Immunity in Multiple Sclerosis: The Dual Signal Hypothesis." *Journal of Neuroimmunolgy* 109, no. 1 (September 1, 2000), 3–9.

Mir, Z. "Clinical Trial at the Klinik Dr. Evers, Study on the Symptomatical Effects of LDN in Multiple Sclerosis." Power Point Presentation, First Annual LDN Conference, New York City, April 2005.

Patel, Priti. "Low-Dose Naltrexone for Treatment of Multiple Sclerosis: Clinical Trials are Needed." Letter to the Editor, *Annals of Pharmacotherapy*, September 2007.

Stambuk, N., et al. "Peptid-M (LUPEX) Immunotherapy in Multiple Sclerosis, Optic Neuritis and Uveitis." *International Journal of Thymology* 5 (1997), 448–464.

Stambuk, N., V. Brinar, Z. Brzovic, et al. "Met-enkephalin Therapy for Autoimmune Diseases: Selective Immunomodulation and Extention of Steroid Therapy." *Journal of Autoimmunity*, Sppl. *Abstracts of 2nd International Congress of Autoimmunity*, edited by J. F. Bach. London: Academic Press, 1999, 86.

Neurodegenerative Disorders Reports and Research

Agrawal, Yash P. "Possible Importance of Antibiotics and Naltrexone in Neurodegenerative Disease." Letter to the Editor, *European Journal of Neurology* 13, no. 9 (2006), e7.

Gilgun-Sherki, Y., M. Hellmann, et al. "The Role of Neurotransmitters and Neuropeptides in Parkinson's Disease: Implications for Therapy. *Drugs of the Future* 29, no. 12 (2004), 1261.

Hu, S., W. S. Sheng, J. R. Lokensgard, and P. K. Peterson. "Morphine Induces Apoptosis of Human Microglia and Neurons." *Neuropharmacology* 42, no. 6 (May 2002), 829–36.

Lipton, S., Z. Gu, and T. Nakamura. "Inflammatory Mediators Leading to Protein Misfolding and Uncompetitive/fast Off-rate Drug Therapy for Neurodegenerative Disorders. *International Reviews in Neurobiology* 82 (2007), 1–27.

Liu, B., L. Du, and J. S. Hong. "Naloxone Protects Rat Dopaminergic Neurons against Inflammatory Damage through Inhibition of Microglia Activation and Superoxide Generation." *Journal of Pharmacology and Experimental Therapeutics* 293 no. 2 (May 2000), 607–617.

_____, H. M. Gao, J. Wang, G. Jeohn, C. Cooper, and Jau-Shyong Hong. "Role of Nitric Oxide in Inflammation-mediated Neurodegeneration. *Annals of the New York Academy of Science* 962, no. 1 (May 2002), 318–33.

_____, and Jau-Shyong Hong. "Role of Microglia in Inflammation-Mediated Neurodegenerative Diseases: Mechanisms and Strategies for Therapeutic Intervention." *Journal of Pharmacology and Experimental Therapeutics* 304, no. 1 (2003), 272.

Rascol, Olivier, Nelly Fabre, Olivier Blin, et al. "Naltrexone, an Opiate Antagonist, Fails to Modify Motor Symptoms in Patients with Parkinson's Disease." *Movement Disorders* 9 no. 4 (October 2004), 437–440.

Salzet, M. "Neuroimmunology of Opioids from Invertebrates to Human." *Neuro Endocrinology Letter* 22, no. 6 (December 2001), 467–74. Review.

Zhang, Wei, Jau-Shyong Hong, Hyoung-Chun Kim, Wanqin Zhang, and Michelle Block. "Morphinan Neuroprotection: New Insight into the Therapy of Neurodegeneration." *Critical Reviews in Neurobiology* 16, no. 4 (2004), 271–302.

Neuroimmunology Reports and Research

Blalock, J. E. "Beta-endorphin in immune cells." *Immunology Today* 19, no. 4 (April 1998), 191–2.

Carr, D. J., K. L. Bost, and J. E. Blalock. "The Production of Antibodies which Recognize Opiate Receptors on Murine Leukocytes." *Life Sciences* 42, no. 25 (1988), 2615–24.

_____, B. R. DeCosta, and A. E. Jacobson, et al. "Corticotropin-releasing Hormone Augments Natural Killer Cell Activity through a Naloxone-sensitive Pathway." *Journal of Neuroimmunology* 28, no.1 (June1990), 53–61.

_____, B. R. DeCosta, C. H. Kim, et al. "Anti-opioid Receptor Antibody Recognition of a Binding Site on Brain and Leukocyte Opioid Receptors." *Neuroendocrinology* 51, no. 5 (May 1990), 552–60.

Carr, D. J., R. T. Radulescu, B. R. DeCosta, K. C. Rice, and J. E. Blalock. "Differential Effect of Opioids on Immunoglobulin Production by Lymphocytes Isolated from Peyer's Patches and Spleen." *Life Sciences* 47, no. 12 (1990), 1059–69.

Faith, R. E., and A. J. Murgo. "Inhibition of Pulmonary Metastases and Enhancement of Natural Killer Cell Activity by Methionine-enkephalin." *Brain, Behavior, and Immunity* 2, no. 2 (June 1988),114–22.

Franco, Rafael, Rodrigo Pacheco, Carmen Liuis, Gerard Ahern, and Petra O' Connell. "The Emergence of Neurotransmitters as Immune Modulators." *Trends in Immunology* 28, no. 9 (September 2007). 400–407.

Heagy, W., E. Teng, et al. "Enkephalin Receptors and Receptor-mediated Signal Transduction in Cultured Human Lymphocytes." *Cellular Immunology* 191, no. 1 (January 10, 1999), 34–48.

Hu, S., W. S. Sheng, J. R. Lokensgard, and P. K. Peterson. "Morphine Induces Apoptosis of Human Microglia and Neurons." *Neuropharmacology* 42, no. 6 (May 2002), 829–36.

Johnson, H. M., E. M. Smith, B. A. Torres, and J. E. Blalock. "Regulation of the In Vitro Antibody Response by Neuroendocrine Hormones." *Proceedings of the National Academy of Sciences USA* 79, no. 13 (July 1982), 4171–4.

"LowDose Naltrexone: Treatment of Endorphin Deficiency," *Fertility Care*, http://www.fertilitycare.net/documents/LDNInfo_000.pdf (accessed May 13, 2008).

Mousa, S. A., R. H. Straub, M. Schäfer, and C. Stein. "Beta-endorphin, Met-enkephalin and Corresponding Opioid Receptors within Synovium of Patients with Joint Trauma, Osteoarthritis and Rheumatoid Arthritis." *Annals of the Rheumatic Diseases* 66, no.7 (July 2007), 871–9.

Olson, G. A., R. D. Olson,, and A. J. Kastin. "Endogenous opiates." *Peptides* 18 (1996), 1651–1688.

Seo, Y. J., M. S. Kwon, S. S. Choi, et al. "Characterization of the Hypothalamic Proopiomelanocortin Gene and Beta-endorphin Expression in the Hypothalamic Arcuate Nucleus of Mice Elicited by Inflammatory Pain." *Neuroscience* 152, no. 4 (2008), 1054–1066.

Shavit, Yehuda, Antoine Depaulis, Fredricka Martin, Gregory Terman, Robert Pechnick, Cynthia Zane, Robert Gale, and John Liebeskind. "Involvement of Brain Opiate Receptors in the Immune-suppressive Effect of Morphine." *Proceedings of the National Academy of Science* 83 (September 1986), 7114–7117.

Stambuk, Nikola. "Cytogenetic Effects of Met-enkephalin (Peptid-M) on Human Lymphocytes." *Croatica Chemica Acta* 71 (1998), 591–605, précis: http://public.carnet.hr/ccacaa/591.html (accessed May 20, 2008).

Terenius, L. "Endorphins and Modulation of Pain." *Advances in Neurology* 33 (1982), 59–64.

Wybran, J., T. Appelboom, J. P. Famaey, and A. Govaerts. "Suggestive Evidence for Receptors for Morphine and Methionine-enkephalin on Normal Human Blood T Lymphocytes." *Journal of Immunology* 123, no.3 (September 1979), 1068–70.

Zagon, I. S., E. Zagon, and P. J. McLaughlin. "Opioids and the Developing Organism: A Comprehensive Bibliography, 1984–1988." *Neuroscience & Biobehavioral Reviews* 13 (1989), 207–235.

Wound Healing and Infections Reports and Research

Kiani, Samira, Mohammad Ebrahimkhani, Ahmad Shariftabrizi, Behzad Doratotaj, et al. "Opioid System Blockade Decreases Collagenase Activity and Improves Liver Injury in a Rat Model of Cholestasis." *Journal of Gastroenterology and Hepatology* 22, no.3 (March 2007), 406–413.

Zagon, I. S. [Ian], J. Jenkins, J. Sassani, J. Wylie, T. Ruth, J. Fry, C. Lang, and P. McLaughlin. "Naltrexone, an Opioid Antagonist, Facilitates Reepitheliazation of the Cornea in Diabetic Rat." *Diabetes* 51 (October 2002), 3055–62.

Zagon, Ian, Matthew Klocek, Joseph Sassani, David Mauger, and Patricia McLaughlin. "Corneal Safety of Topically Applied Naltrexone." *Journal of Ocular Pharmacology and Therapeutics* 22, no. 5 (2006), 377–387.

_____, Joseph Sassani, and Patricia McLaughlin. "Insulin Treatment Ameliorates Impaired Corneal Reepitheliazation in Diabetic Rats." *Diabetes* 55 (April 2006), 1141–47.

Universities and Research Institutions Investigating Low Dose Naltrexone

See Appendix A for details on clinical trials involving LDN, OGF, and other opiate antagonists.

Autism

Bowling Green State University, Department of Psychology, Jaak Panksepp, http:// www.bgsu.edu/departments/psych (accessed May 21, 2008)
Jerusalem Institute for Child Development, Hadassah Medical Organization, David S. Wilensky, http://clinicaltrials.gov/show/NCT00318162 (accessed May 21, 2008)
Schneider Children's Hospital, New Hyde Park, New York, Department of Pharmacy, Gladys Eichaar, Nicole Maisch, Laura Augosto, Heidi Wehrig

Cancer

Pennsylvania State University, College of Medicine, Hershey Medical Center, Ian S. Zagon, http://fred.psu.edu/ds/retrieve/fred/investigator/isz1 (accessed May 21, 2008)

Crohn's Disease

Pennsylvania State University, College of Medicine, Hershey Medical Center, Jill P. Smith, http://www.hmc.psu.edu/colorectal/research/naltrexone.htm (accessed May 21, 2008)

Fibromyalgia

Stanford University School of Medicine, Pain Management Center, Jarred Younger, http://snapl.stanford.edu/ldn (accessed May 21, 2008)

HIV/AIDS

University Hospital in Bamako, Mali, Jaquelyn McCandless, http://www.ldnafricaaids. org

Multiple Sclerosis

University of California, San Francisco, Multiple Sclerosis Research Center, Bruce Cree, http://ucsf.edu/msc/research.htm (accessed May 21, 2008); resources for MS patients: http://www.ucsf.edu/msc/resources.htm (accessed May 21, 2008)
Pennsylvania State University, College of Medicine, Hershey Medical Center, Ian S. Zagon, http://fred.psu.edu/ds/retrieve/fred/investigator/isz1 (accessed May 21, 2008)
San Raffaele Del Monte Tabor Foundation, Maira Gironi, http://www.sanraffaele.org/ EN_home/index.html (accessed May 21, 2008)
The Mind Brain Consortium, David Pincus, http://www.mindbrainconsortium.com (accessed May 21, 2008)

Neurodegenerative Diseases

National Institute of Environmental Health Sciences (NIEHS), Neuropharmacology Group, Jau-Shyong Hong, http://www.niehs.nih.gov/research/atniehs/labs/lpc/neuropharm/index.cfm (accessed May 21, 2008)

Opioid Induced Constipation

Progenics Pharmaceuticals, http://www.progenics.com, http://www.clinicaltrials.gov/ct2/show/NCT00402038 (accessed May 21, 2008); medication: subcutaneous methylnaltrexone

University of Chicago, Pritzker School of Medicine, Chun-Su Yuan, http://www.uchospitals.edu/physicians/physician.html?id=YUAN (accessed May 21, 2008); medication: low dose ethylnaltrexone

Online LDN Resources (English Language)

General Information on LDN (All Conditions)

Clinical Trials for LDN, www.ldninfo.org/ldn_trials.htm (accessed May 21, 2008)

"The Corporate, Political, and Scientific History of Naltrexone," http://www.gazorpa.com/History.html (accessed May 21, 2008)

"How to Talk to Your Doctor About LDN," http://www.gazorpa.com/PatientGuide.html (accessed May 21, 2008)

LDN Annual Conferences, http://ldninfo.org/events.htm (accessed May 21, 2008)

LDN, Low Dose Naltrexone (summary of all conditions for which LDN is being used), http://www.digitalnaturopath.com/treat/T74481.html (accessed May 21, 2008)

Low Dose Naltrexone, http://www.ldninfo.org (accessed May 21, 2008)

Low Dose Naltrexone LDN, by Jeffrey Dach, M.D., http://www.jeffreydach.com/2007/09/01/low-dose-nalotrexone-ldr-by-jeffrey-dach-md.aspx (accessed May 21, 2008)

Low Dose Naltrexone, Wikipedia, http://www.en.wikipedia.org/wiki/Low_dose_naltrexone (accessed May 21, 2008)

"What is Naltrexone?" (includes patient ratings of Naltrexone compared to other drugs, for various conditions; formerly Remedyfind.com), http://www.revolutionhealth.com/drugs-treatments/naltrexone

"An Introduction to Low Dose Naltrexone (LDN)," http://www.webspawner.com/users/introtoldn (accessed May 21, 2008)

Autism

"Naltrexone and Autism," by Stephen M. Edelson, Autism Research Institute, http://www.autism.org/naltrex.html (accessed May 21, 2008)

Autoimmune Disorders

"Low Dose Naltrexone," by Ron Kennedy, The Doctor's Medical Library, http://www.med-library.net/content/view/533/41 (accessed May 21, 2008)

"Low-dose Naltrexone Fights Cancer, HIV, Autoimmune Diseases," Life Extension Foundation forum, http://forum.lef.org/default.aspx?f=37&m=16685 (accessed May 21, 2008)

"Autoimmune Disease," articles by Elaine Moore, http://autoimmunedisease.suite
101.com (accessed May 21, 2008)

Cancer

"LDN and Cancer," http://www.ldninfo.org/ldn_and_cancer.htm (accessed May 21,
2008)
"Low Dose Naltrexone" (includes patient reports and research papers on LDN and cancer), http://www.ldn4cancer.com

Chronic Fatigue Syndrome and Fibromyalgia

"Derek Enlander, M.D., on the Treatment of Chronic Fatigue Syndrome and Fibromyalgia," ProHealth, www.immunesupport.com/library/showarticle.cfm/id/3855 (accessed May 21, 2008)

HIV/AIDS

"Preventing AIDS with LDN in Mali, Africa," http://www.ldnafricaaids.org (accessed
May 21, 2008)
"LDN and HIV/AIDS," www.lowdosenaltrexone.org/ldn_and_hiv.htm (accessed May
21, 2008)
"Low Dose Naltrexone in the Treatment of Acquired Immune Deficiency Syndrome,"
www.lowdosenaltrexone.org/ldn_aids_1988.htm (accessed May 21, 2008)

Multiple Sclerosis

Accelerated Cure Project for Multiple Sclerosis, http://www.acceleratedcure.org (accessed May 21, 2008)
"LDN and Multiple Sclerosis (MS)," http://www.ldninfo.org/ldn_and_ms.htm (accessed May 21, 2008)
LDN Research Trust, www.ldnresearchtrust.org (accessed May 21, 2008)
"Low Dose Naltrexone," news broadcast (WTEV Channel 47 in Jacksonville, Florida),
about an MS patient whose symptoms have eased after using LDN; http://www.you
tube.com/watch?v=Kz52KK5IhOc (accessed May 21, 2008)
Low Dose Naltrexone for Multiple Sclerosis (news, resources and LDN user surveys,
published by SammyJo Wilkinson), http://www.LDNers.org (accessed May 21, 2008)
Multiple Sclerosis Resource Centre, http:/www.msrc.co.uk/index.cfm?fuseaction=show
&pageid=777 (accessed May 21, 2008)
This is MS (includes LDS forum), http://www.thisisms.com (accessed May 21, 2008)

Thyroid Conditions

Alternative Health Solutions for Thyroid Autoimmunity, http://www.ahsta.com (accessed May 21, 2008)

International LDN Information

Polska Strona LDN (Polish-language LDN site), http://www.ldn.org.pl (accessed May
21, 2008)

LDN bei MS Forum (German-language LDN site), http://f27.parsimony.net/forum
67727 (accessed May 21, 2008)

LDN User Stories and Blogs

All Conditions

Those Who Suffer Much Know Much (a publication of Case Health, Brisbane, Queens-
land, Australia, 2001), http://casehealth.com.au./case/pdf/THOSE_WHO_SUFFER_
MUCH_LDN_BOOK_Jul08.pdf.
LDN Research Trust Forum, http://www.ldnresearchtrust.org/forums/. See "Tell Your
Story Here."

Cancer

"How Low Dose Naltrexone Saved Our Lives" (Susie Sedlock's account of using LDN
for multiple myeloma and chronic fatigue syndrome), http://www.susiemaui.com/
low_dose_naltrexone.htm (accessed May 21, 2008)
"My LDN Story," http://www.ldn4cancer.com/myldnstory.html (accessed May 21, 2008)

Multiple Sclerosis

Crystal's MS, Transverse Myelitis and LDN Website, www.freewebs.com/crystal
angel6267 (accessed May 21, 2008)
LDN for MS, http://ldnformultiplesclerosis.blogspot.com (accessed May 21, 2008)
Low Dose Naltrexone aka LDN, http://www.friendswithms.com/low_dose_naltrexone.
htm (accessed May 21, 2008)
"My Story" (SammyJo Wilkonson), http://www.ldners.org/mission.htm (accessed May
21, 2008)
The Mystery Journey—Multiple Sclerosis, http://www.freewebs.com/lovelaugh (accessed May 21, 2008)

LDN Discussion Forums

All Conditions

BrainTalk Communities (use search function for "low dose naltrexone" or LDN),
brain.hastypastry.net/forums (accessed May 21, 2008)
LDN Support, groups.yahoo.com/group/ldnsupport (accessed May 21, 2008)
LDN_Users, groups.yahoo.com/group/LDN_Users (accessed May 21, 2008)
LowDoseNaltrexone (discussion group with over 4,600 members), groups.yahoo
.com/group/lowdosenaltrexone (accessed May 21, 2008)

Autism

Autism_LDN, groups.yahoo.com/group/Autism_LDN (accessed May 21, 2008)

Cancer

LDNforCancer (e-mail group moderated by Dudley Delany), groups.yahoo.com/group/ldnforcancer (accessed May 21, 2008)

HIV, AIDS

LDN in HIV+AIDS, groups.yahoo.com/group/LDN_HIVAIDS (accessed May 21, 2008)

Multiple Sclerosis

Australian Multiple Sclerosis Peer Support, Ozms, http://www.ozms.org (accessed May 21, 2008)
Histamine, LDN & Alternative MS Therapies, http://disc.server.com/Indices/148285.html (accessed May 21, 2008)
Low Dose Naltrexone Forum, http://dn.proboards3.com/index.cgi (accessed May 21, 2008)
Low Dose Naltrexone Forum, This Is MS, http://www.thisisms.com/forum-10.html (accessed May 21, 2008)
MSWatchers, groups.yahoo.com/group/MSWatchers (accessed May 21, 2008)
Spotlight LDN for MS, groups.yahoo.com/group/Spotlight_ldn (accessed May 21, 2008)

LDN Compounding
Pharmacy Information

International Academy of Compounding Pharmacists, http://www.iacprx.org (accessed May 21, 2008)
"LDN Q&A With Dr. Skip," http://www.skipspharmacy.com/ldn.php (accessed May 21, 2008)
Multiple Sclerosis—Naltrexone Information, The Compounder, http://www.thecompounder.com/msnaltrexone.php (accessed May 21, 2008)

COMPOUNDING PHARMACIES KNOWN
TO BE RELIABLE LDN SUPPLIERS

Pharmacy	Phone	Fax
The Apothecary, UT, OH, AZ, NV; mail-order shipping available	800-330-2360	602-277-3148
The Compounder Pharmacy, Aurora, IL	630-859-0333 800-679-4667	630-859-0114
Gideon's Drugs, New York City	212-575-6868	212-575-6334
Grandpa's Compounding Pharmacy, Placerville, CA	530-622-2323	530-622-2011
Irmat Pharmacy, New York City; mail-order shipping available	212-685-0500 800-975-2809	212-532-6596
Maayan Haim, Beit Dagan, Israel	972-(0)3-960-1993	972-(0)3-542-6628
The Medicine Shoppe, Canandaigua, NY	585-396-9970	585-396-7264 800-396-9970

Pharmacy	Phone	Fax
Skip's Pharmacy, Boca Raton, FL	561-218-0111	561-218-8873
	800-553-7429	
Smith's Pharmacy, Toronto, Canada	416-488-2600	416-484-8855
	800-361-6624	

IMPORTANT: *Make sure to specify that you do NOT want LDN in a timed-release or slow-release (SR) form.*

LDN Patient Advice for Doctor Appointments

In addition to the other LDN resources, medical reports and clinical trials described in this book, patients may find the following articles useful to show their doctors, when requesting an LDN prescription:

Clinical Trials for LDN, http://ldninfo.org/ldn_trials.htm (accessed May 21, 2008)
"Interview with LDN Researcher Dr. Agrawal," http://www.bostoncure.org:8080/article.pl?sid=05/01/27/1748256 (accessed May 21, 2008)
"Low Dose Naltrexone," by Elaine Moore, http://autoimmunedisease.suite101.com/print_article.cfm/low_dose_naltrexone (accessed May 21, 2008)
Low Dose Naltrexone Clinical Trials, http://ldners.org/research.htm (accessed May 21, 2008)
"Low Dose Naltrexone Therapy in Multiple Sclerposis," by Yash Agrawal, http://ldners.org/Articles/LDN_Medical_Hypotheses.pdf (accessed May 21, 2008)

Opioid Growth Factor Resources

MedInsight Research Institute, http://www.medinsight.org (accessed May 21, 2008).

Sources of OGF

1) Biofactor GMBH in Germany, http://www.biofactor.de (accessed May 21, 2008). OGF is sold under the brand name Lupex, intended for human use in cancer, AIDS, and autoimmune diseases. This material is not cGMP grade, and therefore is only suitable for subcutaneous injection, not for intravenous infusion. E-mail: info@biofactor.de; telephone: +49 5322 96 0514; fax +49 5322 30 17

2) Netzah Israel Pharmacy in Tel Aviv, Israel, http://www.medinisrael.com (accessed May 21, 2008). Netzah Israel Pharmacy OGF is cGMP grade, and therefore suitable for both subcutaneous and intravenous administration. E-mail: pharmacy@medinisrael.com

Book Sources on Opioid Growth Factor

Loughlin, Sandra E., and James H. Fallon. *Neurotrophic Factors*. San Diego, CA: Academic Press, 1993.
Morstyn, George, MaryAnn Foote, and Graham J. Lieschke. *Hematopoietic Growth Factors in Oncology: Basic Science and Clinical Therapeutics*. Totowa, NJ: Humana Press, 2004.

Restorative Therapies
for Patients with
Chronic Diseases

Chiropractic

Chiropractic therapy is intended to restores normal function of the nervous system by manipulation and treatment. It is usually covered by insurance and Medicare; check with your provider.

Find a practitioner: American Chiropractic Association, http://www.amerchiro.org/search/memsearch.cfm (accessed May 21, 2008).

CranioSacral Therapy

Craniosacral therapy is a gentle manipulative technique, performed by a trained practitioner, intended to find and correct cerebral and spinal imbalances or blockages that may cause tissue, emotional and postural dysfunction.

Find a practitioner: International Association of Healthcare Practitioners, http://www.iahp.com/pages/search (accessed May 21, 2008).

Massage

Medical massage is available at most hospitals; deep tissue massages are also available at many spas. Massage helps increase circulation to muscles and helps relax tense, knotted muscles. This does not require a prescription and is occasionally covered by insurance; check with your provider. Look for certified medical massage therapists.

Find a practitioner: American Massage Therapy Association, http://www.amtamassage.org/findamassage/locator.aspx (accessed May 21, 2008).

Physical therapy
or Physiotherapy

Health care concerned with prevention and management of movement disorders arising from conditions or diseases, through the use of guided exercises to help condition muscles and restore strength, movement and neurologic function, typically requires a prescription from your doctor. It is usually covered by insurance and Medicare; check with your provider.

Feldenkrais

Feldenkrais is a form of neuro-motor reconditioning by "Awareness Through Movement" classes and at-home exercise, as well as gentle manipulative techniques by a trained practitioner that promote recovery from neurodegenerative and physically wasting conditions.

Find a practitioner: The Feldenkrais Method of Somatic Education, http://www.feldenkrais.com/practitioners/find (accessed May 21, 2008).

General Health
Information Resources

U. S. Government Health Information and Medical Research

Consumer Health Information in Many Languages Resources, www.nnlm.gov/out-
reach/consumer/multi.html (accessed May 22, 2008)
This National Network of Libraries of Medicine site includes links to medical information in Arabic, Cambodian/Khmer, Chinese, French, German, Hmong, Korean, Laot-
ian, Russian, Somali, Spanish, Thai, and Vietnamese.
DIRLINE, Directory of Health Organizations, http://dirline.nlm.nih.gov (accessed May
22, 2008)
Compiled by the National Library of Medicine, this site contains descriptive informa-
tion about a variety of health organizations.
Drug Information Portal, U.S. National Library of Medicine, http://druginfo.
nlm.nih.gov (accessed May 22, 2008)
The NLM Drug Information Portal gives users a gateway to selected drug information.
"Guidance for Institutional Review Boards and Clinical Investigators," 1998 Update,
U.S. Food and Drug Administration, http://www.fda.gov/oc/ohrt/irbs/offlabel.html
(accessed May 22, 2008)
Health.gov, http://www.health.gov (accessed May 22, 2008) "A portal to the Web sites
of a number of multi-agency health initiatives and activities of the U.S. Department
of Health and Human Services (HHS) and other Federal departments and agencies."
Healthfinder, http://www.healthfinder.gov (accessed May 22, 2008) The Healthfinder
project is coordinated by the U.S. Department of Health and Human Services Office
of Disease Prevention and Health Promotion. It is a resource for finding the best government and nonprofit health and human services information on the Internet.
Health for Seniors, http://www.usa.gov/Topics/Seniors/Health.shtml (accessed May 22,
2008)
This site is a portal to official information and services from the U.S. government on
diseases, prescription drugs, Medicare and Medicaid, and doctors and health-care facilities for seniors.
MEDLINEplus, http://www.medlineplus.gov (accessed May 22, 2008) From the National Library of Medicine, MEDLINEplus has extensive information on over 500 diseases and conditions. There are also lists of hospitals and physicians, a medical en-
cyclopedia and dictionaries, health information in Spanish, extensive information on
prescription and nonprescription drugs, and links to thousands of clinical trials.
National Center for Complementary and Alternative Medicine (NCCAM), http://
www.nccam.nih.gov (accessed May 22, 2008)
The NCCAM site includes research results and information about clinical trials.
PubMed.gov, http://www.ncbi.nlm.nih.gov/pubmed (accessed May 22, 2008)
PubMed is a free digital archive of biomedical and life sciences journal literature.
Tox Town, http://www.toxtown.nlm.nih.gov (accessed May 22, 2008)
This National Library of Medicine site is an introduction to toxic chemicals and en-
vironmental health risks you might encounter in everyday life.

U.S. University Health Information Services

Emory University, MedWeb Community, http://www.medweb.emory.edu/SPT—
Home.php (accessed May 22, 2008)

MedWeb is "a catalog of biomedical and health related web sites maintained by the staff of the Robert W. Woodruff Health Sciences Center Library at Emory University."
Harvard Medical School Health Information & Medical Information, http://www.health.harvard.edu (accessed May 22, 2008)
Mayo Clinic, http://www.mayoclinic.com (accessed May 22, 2008)
Includes information on specific diseases and conditions, and prescription and over-the-counter drugs, as well as interactive tools such as exercise and stress management planners.
NetWellness, Consumer Health Information, http://www.netwellness.org (accessed May 22, 2008)
This nonprofit site, in operation for over ten years, provides information "created and evaluated by medical and health professional faculty at the University of Cincinnati, Case Western Reserve University, and Ohio State University.
Tufts University Health & Nutrition Letter, http://www.healthletter.tufts.edu (accessed May 22, 2008)
The content of this newsletter is "based substantially on the research and expertise of the Gerald J. and Dorothy R. Friedman School of Nutrition Science and Policy."
University of California, San Francisco, Multiple Sclerosis Center, http:// ucsf.edu/msc /resources.htm (accessed May 22, 2008)
This page provides extensive links to MS and disability organizations and services.
Yale School of Public Health, Consumer Health, http://www.info.med.yale.edu/ library/consumer (accessed May 22, 2008)
This Yale site is a portal to consumer health information resources.
Yale School of Public Health Library, www.info.med.yale.edu/eph/phlibrary (accessed May 22, 2008)
This site provides links to medical research databases.

International Health Information Services

Croatian Scientific Bibliography (CROSBI), http://bib.irb.hr/index.html?lang=EN (accessed May 22, 2008)
CROSBI "stores scientific papers published in the period from 1997 to the present," covering all scientific publications in Croatia (some in English).

Health Information Portals (Nongovernmental, Nonacademic)

Aetna InteliHealth, http://www.intelihealth.com (accessed May 22, 2008)
Partnering with the Harvard Medical School, InteliHealth includes health news and access to "interactive tools" and risk assessments.
American Association of Retired Persons, Health, http://www.aarp.org/health (accessed May 22, 2008)
This site is a source of information on wellness and fitness for people over fifty.
Familydoctor.org, http://www.familydoctor.org (accessed May 22, 2008)
Famildoctor.org offers information written and reviewed by physicians and patient education professionals at the American Academy of Family Physicians. Topics include health and wellness, drug information, herbal and alternative remedies, and self-care.
Medicine Chest, Revolution Health, http://www.revolutionhealth.com/drugs-treatments (accessed May 22, 2008)
Includes patient ratings on medications including naltrexone (http://www.revolutionhealth.com/drugs-treatments/rating/naltrexone).

PatientsLikeMe.com, http://www.patientslikeme.com (accessed May 22, 2008)
   On this interactive site, MS patients using LDN report their progress, with graphs and ratings (http://www.patientslikeme.com/treatments/show/64).
WebMD, http://www.webmd.com (accessed May 22, 2008)
   This site offers a variety of health and medical news and information.

National Health Organizations and Resources

These organizations that focus on patient assistance as well as research.
American Parkinson Disease Association, http://www.apdaparkinson.org (accessed May 22, 2008) Phone: 800-223-2732
Autism Research Institute, http://www.autism.com (accessed May 22, 2008) Phone: 866-366-3361
Autoimmune Information Network, http://www.aininc.org (accessed May 22, 2008); Phone: 877-246-4900
Crohn's and Colitis Foundation of America, http://www.ccfa.org (accessed May 22, 2008); Phone: 800-932-2423
Gulf War Syndrome forum, http://www.gulfweb.org/fusetalk/categories.cfm?catid=7 (accessed May 22, 2008)
Multiple Sclerosis Foundation, http://www.msfocus.org (accessed May 22, 2008); Phone: 888-MS-FOCUS ("helpline" staffed by caseworkers and peer counselors)
Multiple Sclerosis International Federation (MSIF), http://www.msif.org (accessed May 22, 2008)
International body linking national MS societies around the world. Directory of local MS societies, research news
National Cancer Institute, http://www.cancer.gov (accessed May 22, 2008)
Comprehensive cancer information from the U.S. government's principal agency for cancer research; Phone: 800-422-6237
National Chronic Fatigue Syndrome and Fibromyalgia Association (NCFSFA), http://www.ncfsfa.org (accessed May 22, 2008); Phone: 816-737-1343
National Fibromyalgia Association, http://www.fmaware.org (accessed May 22, 2008); Phone: 714-921-0150
National Multiple Sclerosis Society, http://www.nationalmssociety.org (accessed May 22, 2008); Phone: 800-344-4867
Thyroid Foundation of America, http://www.tsh.org (accessed May 22, 2008) Phone: 800-832-8321

Alternative Health Resources

The Alternative Medicine Homepage, http://www.pitt.edu/~cbw/altm.html (accessed May 22, 2008) This site, a portal for "sources of information on unconventional, unorthodox, unproven, or alternative, complementary, innovative, integrative therapies," is maintained by medical librarian, Charles B. Wessel, at the University of Pittsburgh.
CPNhelp.org, http://www.CPNhelp.org (accessed May 22, 2008) This site is devoted to the understanding and treatment of *Chlamydia pneumoniae*, a species of infectious bacteria implicated in a number of human illnesses, including multiple sclerosis, chronic fatigue, cardiac disease, interstitial cystitis, prostatitis, Crohn's disease, inflammatory bowel disease, Alzheimer's disease, asthma, arthritis, fibromyalgia, chronic refractory sinusitis, and macular degeneration.
HonestMedicine.com, http://www.honestmedicine.com (accessed May 22, 2008) Julia

Schopick, who says that she is "committed to working with others to make significant changes to the way people think about, and interact with, the medical system," uses her Web site to spotlight new (and often not-yet-accepted) medical ideas and treatments.

"Hughes Syndrome," Multiple Sclerosis Resource Centre, http://www.msrc.co.uk/index.cfm/fuseaction/show/pageid/736 (accessed May 22, 2008) This page features articles and information about "the blood disease that mimics MS—Hughes Syndrome."

"Lyme Disease Misdiagnosed as Multiple Sclerosis," LymeInfo.net, http://www.lymeinfo.net/multiplesclerosis.html (accessed May 22, 2008) This page links to abstracts that provide information about Lyme disease being misdiagnosed as MS.

Office of Cancer Complementary and Alternative Medicine (OCCAM), National Health Institute, http://www.cancer.gov/cam (accessed May 22, 2008) The OCCAM site includes information about clinical trials.

SavvyPatients.com, Integrative Medicine on the Internet, http://www.savvypatients.com (accessed May 22, 2008) This site offers information about alternative approaches to many different diseases, and includes links to more information about LDN.

Rocky Mountain MS Center, http://www.ms-cam.org (accessed May 22, 2008) The site offers information on complementary and alternative medicine (CAM) and multiple sclerosis.

Clinical Trials Directories

Accelerated Cure Project for Multiple Sclerosis, Clinical Trials Resource Center, http://www.acceleratedcure.org/msresources/trials.php (accessed May 22, 2008) This site includes access to information about "U.S. and international clinical trials actively recruiting patients for multiple sclerosis."

CenterWatch Clinical Trials Listing Service, http://www.centerwatch.com (accessed May 22, 2008) You can use this site to find "information about clinical research, including over 25,000 clinical trial listings (both industry and government sponsored), new drug therapies in research and those recently approved by the FDA." It's both for patients interested in participating in clinical trials and for research professionals.

ClinicalTrials.gov, http://www.clinicaltrials.gov ClinicalTrials.gov is a searchable " registry of federally and privately supported clinical trials conducted in the United States and around the world."

Clinical Trials for LDN, ldinfo.org, http://www.ldninfo.org/ldn_trials.htm (accessed May 22, 2008)

National Cancer Institute, Clinical Trials, http://www.cancer.gov/clinicaltrials (accessed May 22, 2008) This site includes both clinical trial results and a list of trials accepting participants.

National Center for Complementary and Alternative Medicine (NCCAM), Clinical Trials, http://nccam.nih.gov/clinicaltrials (accessed May 22, 2008) This site provides access to information about health care research outside the realm of conventional medicine.

National Multiple Sclerosis Society, http://www.nationalmssociety.org (accessed May 22, 2008) Search for "clinical trials" on this site for information on MS trials and recruitment.

Resources to Help Assess Benefits and Risks of Participating in a Clinical Trial

"Clinical Cancer Trials: The Basic Workbook," National Cancer Institute, http://www.cancer.gov/clinicaltrials/resources/basicworkbook/page3 (accessed May 22, 2008)

"Clinical Trials," U.S. Food and Drug Administration, http://fda.gov/oashi/clinicaltrials (accessed May 22, 2008)

"Drug Safety," MedLine Plus, http://nlm.nih.gov/medlineplus/drugsafety.html (accessed May 22, 2008)

"Should You Participate in Clinical Trials?" The Body, http://thebody.com/content/art12698.html (accessed May 22, 2008)

"Understanding Clinical Trials," ClinicalTrials.gov, http://clinicaltrials.gov/ct2/info/understand (accessed May 22, 2008)

Index